Advances in Wastewater Treatment II

Edited by

Kinjal J. Shah[1] and Vimal Gandhi[2]

[1]College of Urban Construction, Department of Municipal Engineering, Nanjing Tech University (NTU), Nanjing, China 218 816

[2]Department of Chemical Engineering, Faculty of Technology, Dharmsinh Desai University, Nadiad-387 001, Gujarat, India

Published by **Materials Research Forum LLC**
Millersville, PA 17551, USA

Published as part of the book series
Materials Research Foundations
Volume 102 (2021)
ISSN 2471-8890 (Print)
ISSN 2471-8904 (Online)

Print ISBN 978-1-64490-138-0
eBook ISBN 978-1-64490-139-7

Distributed worldwide by

Materials Research Forum LLC
105 Springdale Lane
Millersville, PA 17551
USA
https://www.mrforum.com

Manufactured in the United States of America
10 9 8 7 6 5 4 3 2 1

Table of Contents

Preface

Advances in wastewater treatment is an essential technological aspect wherein purification of industrial and municipal wastewater can be achieved by reuse and recycle technology. It is found that 30-50% of the industrial wastewater flows back into the ecosystem globally either being poorly treated or without being treated or reused depending upon the types of industries which adversely affect the ecosystem and environment. This statistic indicates the importance of water and the need of its treatment technologies for the coming decades. Thus, the United Nations General Assembly prioritize goal 6 as "clean water and sanitation" among the other 17 different goals for achieving overall sustainability. This book provides comprehensive information on advances in wastewater technologies in 7 chapters. We are sure that engineering, scientist, project manager and researcher will consider this book as essential reference tool on advances in wastewater treatment technologies.

In our edited book **Advances in Wastewater treatment I**, the utilization of alternative treatment methods in the collaboration of conventional methods to treat the industrial and municipal wastewater were elaborated. These includes mainly (i) summarized conventional waste water treatment technologies along with need and types of advance waste water treatment technologies; (ii) focusing on review on heterogeneous advance oxidation process; (iii) application of AOPs for degradation of pharmaceutical compounds; (iv) advancement in membrane technology for arsenic removal; (v) development and applicability of low cost adsorbent; (vi) degradation of polycyclic aromatic hydrocarbons by microorganism and powder activated carbon; and (vii) various types of bio-flocculant for wastewater treatment. Still, the need to find alternative advance technologies for effectively treatment of industrial and municipal wastewater is vital. In that context, the current book **Advances in Wastewater Treatment II** summarized some additional technologies.

In Part I, it is found that heterogeneous Advanced Oxidation Processes (AOPs) are gaining importance due to its usefulness to degrade complex organic molecules. In this book, **Chapter 1** is focusing on the review of homogeneous AOPs. **Chapter 2** emphasized on application of one of the AOPs, electrochemical oxidation (EO) of perfluorooctanoic acid (PFOA) using $Ti/SnO_2\text{-}Sb_2O_5/PbO_2$ anodes in aqueous solution.

Membrane based Reverse osmosis, Nano filtration, Ultrafiltration and Microfiltration are gaining importance as a part of advanced wastewater treatment technologies. **Chapter 3** nicely reviews recent developments in the area of membrane-based processes for the treatment of wastewater including challenges involved in the field of membrane technology. Whereas applications of Nanomaterials and nanocomposite have been the new area of research in the field of advance wastewater treatment technologies. **Chapter 4** focused on green and sustainable organic-inorganic metal nanocomposites based on nano carbon cage for the removal of various dyes from wastewater.

Adsorption process is also widely used as a part of tertiary treatment or polishing methods to treat wastewater. Adsorption column designing is one of the important task to convert lab scale data into real industrial application using suitable mathematical model reported in the literature. **Chapter 5** aimed for providing approaches in selecting the best performing model for the design of adsorption column.

Due to rapid population growth, utilization of various pesticide compounds has tremendously increase in farming and other sectors. The higher amounts of pesticides found in municipal and industrial wastewater leads to serious threat to the water ecosystem. Removal and degradation of pesticides from wastewater is considered to be a challenging task. Apart from conventional treatment techniques, newly developed hybrid process of "hydrolytic acidification + two-stage Expanded Granular Sludge Bed (EGSB) + two-stage advance Fenton oxidation" are reported in **Chapter 6**.

Heterogeneous Photocatalysis is an integral part of Advanced Oxidation Processes. The removal of pharmaceutical compounds from waste water is a challenging task in the Covid era. **Chapter 7** emphasized on photocatalytic degradation of Levofloxacin by metal doped titanium dioxide under the visible light source.

In total, both volumes provide remarkable information regarding advancement in wastewater treatment through many outstanding reviews and research article from various distinguished expert from their areas. We are thankful to all the authors for their willingness to provide book chapters and join hands with us to create awareness about advances in wastewater treatment. We would also like to express our gratitude to all the publishers and authors for granting us the copyright permission to reprint the same. Although sincere efforts have been made to get copyright permissions from the respective agencies, we sincerely apologies to any copyright holder if unknowingly their right is being infringed.

We would like to take this opportunity to express our sincere gratitude to KJS lab members of Nanjing Tech University, China and VGG group members of Dharmsinh Desai University-Nadiad, India for their technical support and guidance. Our sincere thanks to Dr. P.A.Joshi and Dr. Meha Sanghvi for their supports in review the book chapters and provided valuable suggestions. Our special thanks go to Mr. Thomas Wohlbier of Materials Research Forum LLC, and his editing team for whole hearted support extended to us from proposal stage to printing stage. We are also thankful to all the friends and family members for their best wishes and blessings.

Dr. Kinjal J. Shah
College of Urban Construction,
Department of Municipal Engineering
Nanjing Tech University (NTU),
Nanjing, China 218 816

Dr. Vimal Gandhi
Department of Chemical Engineering,
Faculty of Technology,
Dharmsinh Desai University,
Nadiad-387 001, Gujarat, INDIA

Advances in Wastewater Treatment II — Materials Research Forum LLC
Materials Research Foundations **102** (2021) 1-47 — https://doi.org/10.21741/9781644901397-1

Chapter 1

Research Progress of Advanced Oxidation Water Treatment Technology

Yongjun Sun*, Deng Li, Shengbao Zhou, Kinjal J. Shah*, Xuefeng Xiao
College of Urban Construction, Nanjing Tech University, Nanjing, 211816, China
sunyongjun@njtech.edu.cn, kjshah@njtech.edu.cn

Abstract

The rapid development of the industry has brought environmental deterioration. The contradiction between industrial development and environmental protection has become increasingly prominent. How to degrade various complex pollutants in water effectively is one of the keys to solve this contradiction. Reasonable treatment and disposal of pollutants to make them harmless to the environment has important practical significance. Advanced oxidation Processes (AOPs) has been gradually applied in the field of water treatment due to its advantages of high treatment efficiency, strong oxidation intensity, no secondary pollution and its capacity to degrade varieties of organic pollutants from wastewater. The oxidative degradation mechanism of AOPs was analyzed in this paper. The future development trend of the advanced oxidation technology was prospected, and the best treatment method for water treatment technologies was sought.

Keywords

Advanced Oxidation Technology, Water Treatment, Oxidation Performance, Oxidation Mechanism

Contents

Materials Research Forum LLC
https://doi.org/10.21741/9781644901397-1

1. Introduction

Sustainable development goals are a bunch of 17 goals announced by the United Nations to achieve global sustainability. Among them, safe and affordable drinking water for all is an alarming situation for major countries due to the issue of water scarcity [1]. After considering the gravity of the problem, the UN has considered clean water and sanitation as an independent goal (Goal 6) among their 17 goals. Meanwhile, the quality of water resources is not only related to the development of industry and agriculture but also plays a decisive role in the sustainable development of the entire national economy and society. On the other hand, the rapid development of industry and agriculture and the rapid increase in population has caused water quality pollution and making serious water resources problems around the world. An important source of water pollution is industrial wastewater [2]. Industrial wastewater often contains a large amount of toxic and refractory organic pollutants, such as polycyclic aromatic hydrocarbons, herbicides, insecticides, and dyes [3], because of its stable and undegradable chemical structure [4]. Thus, it has been remaining in nature for a long time as untreatable, and it has been widely concerned with humans, animals, and aquatic life [5]. In addition, these refractory organic pollutants enter the water body without treatment, they will easily form pollution to surface water, groundwater and soil, and then affect human health through drinking water and the food chain, which pose a serious threat to human sustainable development. For these refractory organic pollutants, it is extremely necessary to treat the wastewater containing these pollutants before entering the water environment.

There are many methods for treating organic pollutants, such as physical methods (physical adsorption [6], extraction [7], distillation [8], flotation [9], flocculation [10] and reverse osmosis [11], etc.), chemical methods (incineration, ozone, micro-electrolysis [12] and wet oxidation, etc.) and biological methods (activated sludge method [13] and biofilm method [14], etc.). The biological treatment method is widely used in wastewater treatment due to its advantages of simple equipment and low operating cost. However, its long treatment cycle, parametric sensitivity and large equipment occupation area, it is difficult to achieve good results by using ordinary biological treatment methods. Traditional physical and chemical methods such as activated carbon adsorption, flocculation, and gas stripping can transfer organic pollutants from the liquid phase to the solid phase or the gas phase and do

Materials Research Forum LLC
https://doi.org/10.21741/9781644901397-1

not eliminate the organic pollutants. Moreover, due to technical or economic reasons, organic pollutants cannot be well recycled and rather creating causing problems for treatment i.e. waste accumulation or secondary pollution. The current time factor that permits the researcher to focus on this field is to seek an efficient and secondary pollution-free treatment method [15].

Advanced Oxidation Processes (AOPs) are a new class of technology for oxidative removal of organic pollutants, which refers to the use of strong oxidizing radicals generated in the reaction as the main oxidant for oxidative decomposition and mineralization of organic matter [2]. Compared with other oxidation methods, AOPs has the following characteristics i.e.

1. It produces a lot of active hydroxyl (·OH) radicals and induces the follow-up chain reactions;

2. ·OH radicals indirectly react with organic pollutants in wastewater to mineralize it to water and CO_2 [16].

3. It can be used as a separate treatment process, or it can be matched with other treatment processes, such as pre- and post-treatment of biochemical treatment, which can reduce the treatment cost.

AOPs can be divided into the Fenton oxidation method, ozone oxidation method, ultrasonic oxidation method, photocatalytic oxidation method, wet oxidation method, supercritical water oxidation method, electrochemical oxidation methods [2, 17]. In addition to the above-advanced oxidation technology based on hydroxyl radical ·OH, there is also a persulfate oxidation method which is based on SO_4-·. Compared with traditional oxidation technology, the AOPs have many advantages such as strong oxidation, low selectivity, fast reaction rate, simple operation, easy control, high efficiency, and wide processing range [18].

Based on recent literature, this article mainly reviews the formation of free radicals generated by various advanced oxidation technologies listed in previous paragraphs. The mechanism summarizes the application in various wastewaters and puts forward the advantages and disadvantages of various oxidation technologies in the treatment of wastewater. The advanced oxidation technologies have different processes for oxidizing organic substances. This article also gives some typical AOPs process and describes the process of degrading organic matter.

Materials Research Forum LLC
https://doi.org/10.21741/9781644901397-1

2. Fenton oxidation technology

2.1 Mechanism of Fenton oxidation

In 1894, French scientist Fenton discovered that under acidic conditions, the addition of divalent iron or divalent copper ions to hydrogen peroxide can effectively oxidize tartaric acid [19]. Later on, in the 1960s, Eisenhower first applied the Fe^{2+}/H_2O_2 system to treat phenolic wastewater and alkyl wastewater. Then after, Fe^{2+}/H_2O_2 widely known as a Fenton reagent. Since then, the Fenton oxidation method has been receiving more and more attention in the research of organic wastewater treatment.

In the study of the Fenton oxidation mechanism, the most classic reaction mechanism is "·OH radical theory". Researchers from Utah State University proposed the possible generation mechanism of free radicals and oxidant fragments during Fenton reaction. After many years of research, the mechanism of Fenton oxidation under H_2O_2, generates strong ·OH and achieve the degradation of organic matter [20]. The oxidation process mainly involves the reaction equations (See equations 1-1 to 1-7). The schematic diagram of the reaction mechanism is shown in Figure 1. The production of ·OH is the beginning of the chain (equation 1-1). The interaction of each active oxygen and active oxygen and other substances causes the consumption of active oxygen and the termination of the reaction chain. These reactive oxygen species react with organic molecules and mineralize them into CO_2 and H_2O (equation 1-7). The ·OH radicals produced in the reaction is a highly reactive with oxidation-reduction potential is 2.8eV [21].

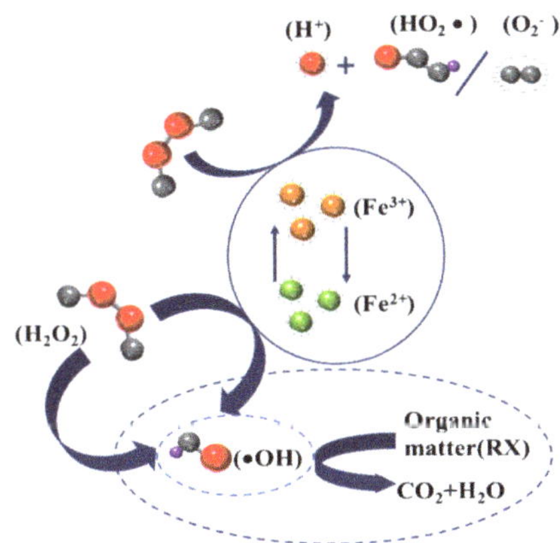

Fig.1 Schematic diagram of the Fenton reaction mechanism.

Materials Research Forum LLC
https://doi.org/10.21741/9781644901397-1

$Fe^{2+}+H_2O_2 \rightarrow Fe^{3+}+\cdot OH+OH^-$ (1-1)
$Fe^{2+}+\cdot OH \rightarrow Fe^{3+}+OH^-$ (1-2)
$Fe^{3+}+H_2O_2 \rightarrow Fe^{2+}+H_2O\cdot+H^+$ (1-3)
$H_2O\cdot+H_2O_2 \rightarrow O_2+H_2O+\cdot OH$ (1-4)
$RH+\cdot OH \rightarrow R\cdot+H_2O$ (1-5)
$R\cdot+Fe^{3+} \rightarrow R^++Fe^{2+}$ (1-6)
$R^++O_2 \rightarrow ROO^+ \rightarrow CO_2+H_2O$ (1-7)

2.2 Progress in application of Fenton technology

Fenton reagent is widely used in the treatment of industrial wastewater. He et al. [22] used Fenton oxidation to treat alkaline scarlet dye wastewater and investigated the effects of pH, reaction time, Fenton reagent ratio, and dosage on the colorimetric treatment effects. He found, when pH = 3, ratio of C (Fe^{2+}): C (H_2O_2) = 1:5, and the reaction time was 30 min, the removal rate of wastewater color was reached to 92.30%. Jia et al. [23] studied that the treatment of wastewater containing methyl orange and methylene blue mixed dye by the Fenton oxidation method. At pH = 2.5, Fe^{2+} dosage of 5 mmol, H_2O_2/Fe^{2+} molar ratio of 4, reaction time of 50 min; the removal rate of chroma, TOC, COD, and ammonia nitrogen were found to be 99.8%, 85.11%, 90.69%, and 66.33%, respectively.

Although Fenton's reagent is considered to be a strong oxidant and used in the treatment of a variety of industrial wastewater, it has its inherent disadvantages and limitations too. The reaction only has better activity when the pH is close to 3. At higher pH, such as pH greater than 6, the effect of Fenton's reagent is greatly reduced due to iron aggregation and sedimentation [24]. Since the pH of most wastewater is greater than 6, it is necessary to adjust the acidity of the wastewater in advance and add acidic reagents, which increases the operating cost. In most cases, due to the easy occurrence of iron aggregation and sedimentation, the intermediate products of degradation can form a stable complex with Fe^{3+}, which reduced the degree of mineralization of organic pollutants. Moreover, it is difficult to generate higher amount of $\cdot OH$ radicals during the presence of carbonate and phosphate in natural water. This behaviour is observed due to quenching and inactivation of Fenton reagent. It has been reported that carbonate in water can be used as a quencher for light-based free radicals, resulting in a decrease in the treatment effect of Fenton reagent.

In order to overcome such barriers, many researchers have made improvements to the Fenton oxidation method, such as introducing ultraviolet light in the conventional Fenton system. Li et al. [25] studied the UV-Fenton oxidation method for the treatment of wine production wastewater. When the optimal reaction conditions were pH 3, reaction time 30 min, and C(Fe^{2+}) = 40 mmol/L, the COD removal rate was reached to be 87%. The advantage of the UV combined Fenton method is to reduce the amount of Fe^{2+} and increases the utilization rate of H_2O_2. This is due to the synergistic effect of Fe^{3+} and ultraviolet light on the catalytic decomposition of H_2O_2. In recent years, people have tried to replace Fe^{2+} in traditional Fenton reagents with Fe^{3+}, but the degradation rate of organic matter by this catalytic system under dark reaction conditions (without light) is much lower than that of traditional Fenton reagents. On the other hand, the same system can greatly accelerate the degradation rate of organic matter under light condition, and the utilization rate of H_2O_2 is greatly improved. The form of Fe^{3+} in aqueous solution is mainly related to the pH of the medium. At a pH value equal to about 3, Fe^{3+} mainly exists in the form of Fe $(OH)^{2+}$. The basified iron appear yellow and show an absorption band of charge transfer between the ligand-center iron atom in the ultraviolet region (the tail is close to the visible region), and $Fe(OH)^{2+}$ of Fe^{3+} absorbs ultraviolet light ·OH and ferrous irons. The Fe^{2+} produced by photoreduction reacts with H_2O_2 again, increasing the ·OH yield, forming a cycle of Fe^{3+} /Fe^{2+}, thereby accelerating the decomposition rate of organic matter.

The conventional improvement of the Fenton oxidation method is still a homogeneous system. There are still problems that are difficult to solve when using the homogeneous Fenton system. For example, the homogeneous system will produce a large amount of iron sludge, so people began to focus on the heterogeneous system. Tang et al. [26] prepared three-dimensional electrodes with stainless steel plates, ruthenium-iridium-plated electrodes as cathode and anode electrodes, and powdered activated carbon as particle electrodes. This three-dimensional electrode was used to treat the actual dye wastewater. When the dosage of powdered activated carbon was 2.0g/L. The current density is 0.5 mA/mm^2, the plate spacing is 3 cm, pH = 2.0, the dosage of ferrous sulfate is 0.5 g/L, and the reaction time is 2 h, the removal rates were found to be 62.80%, 41.15%, 42.48%, and 95.00%, respectively. The heterogeneous system Fenton significantly enhances the oxidative degradation ability of Fenton reagents to organic matter and can reduce the amount of H_2O_2 and reduce the treatment cost. Whether it is a heterogeneous Fenton oxidation method or a homogeneous Fenton oxidation method that introduces ultraviolet light, the basic principles of these technologies are similar to the original Fenton oxidation method. The main oxidation function in the process of treating organic pollutants is ·OH and they are collectively referred to as Fenton-like reactions.

2.3 Development of Fenton technology

A large number of experimental studies have shown that Fenton reagent is a high-efficiency oxidant, suitable for the treatment of refractory organic wastewater. However, the application of Fenton reaction to the field of wastewater treatment and the industrial application still have the following main problems: the treatment cost is relatively high; the generation rate and utilization rate of hydroxyl radicals ·OH are not high; the reaction conditions are strict and it is easy to introduce secondary pollution [27]. Appropriate introduction of new technologies improves the production mechanism of hydroxyl radicals ·OH in Fenton reaction, increases the generation rate of hydroxyl radicals ·OH, and prolongs the existence time of ·OH [28]. At present, improving the production mechanism of hydroxyl radicals ·OH mainly develops along with the two directions of photochemistry and electrochemistry. Through photochemical or electrochemical action [29], it can directly or indirectly promote (decompose H_2O or H_2O_2) the production of hydroxyl radicals ·OH, thereby increasing the generation rate of hydroxyl radicals ·OH; at the same time can also reduce the amount of H_2O_2 and Fe^{2+}, thereby effectively reducing wastewater treatment costs.

Moreover, Fenton technology can be combined with microwave and ultrasonic technology to improve the generation rate and utilization rate of hydroxyl radical ·OH. In addition to improving Fenton oxidation technology in combination with other new technologies, it is also a possibility of improving the Fenton reaction conditions by adding homogeneous and heterogeneous catalysts and explores broadly applicable conditions and environmentally friendly wastewater treatment.

3. Ozone oxidation technology

3.1 Mechanism of oxidation by ozone

In 1840, Schönbein discovered ozone. In 1872, the chemical structure of ozone was finally confirmed as a triatomic oxygen molecule. In late 1886, de Meritens discovered that ozone can be used as a bactericide to sterilize sewage [30]. Subsequently, after several years of pilot tests at a water treatment plant in Paris, ozone was first used in water treatment (and continuously using). Disinfection of drinking water was carried out in France in 1906 [31] and in 1948, people began research on catalytic ozonation technology. In 1972, the homogeneous catalytic ozonation process was first applied to the degradation of pollutants in water. In 1997, Kaptijn first reported the catalytic ozonation reaction using activated carbon as a homogeneous catalyst. Ozone is an excellent strong oxidant and has strong adaptability to oxidize organic or inorganic substances in the water at any pH [32]. Ozone can oxidize most organic substances, especially those that are difficult to degrade and have

Materials Research Forum LLC
https://doi.org/10.21741/9781644901397-1

good effects. During the reaction of ozone with organic matter in water, there are usually two ways of direct reaction and indirect reaction [33]. The schematic diagram of the indirect reaction path of ozone in water is shown in Figure 2.

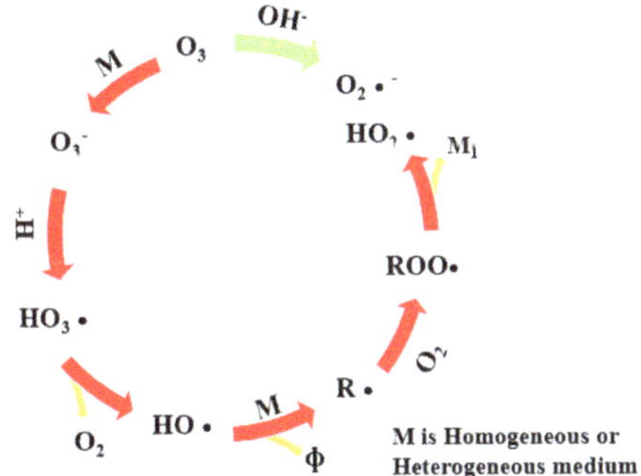

Fig.2 Schematic diagram of the indirect reaction path of ozone in water.

(1) Direct oxidation reaction

Ozone direct reaction is referred as the direct oxidation of the organic matter. In general, direct oxidation reaction is slow and selective, and the reaction rate constant is in the range of $10\text{-}10^3 M^{-1}s^{-1}$. Ozone direct oxidation mechanisms include cycloaddition reaction [34], electrophilic reaction, and nucleophilic reaction. Cycloaddition reaction refers to the fact that O_3 has bipolarity, which makes it easy for O_3 molecules to be added to unsaturated bonds, which breaks the bonds and generates compounds with carbon groups and carboxyl groups, such as carboxylic acids, ketones, and aldehydes. Electrophilic reaction means that on the ortho and para carbon atoms of some aromatic compounds containing electron-donating groups (-OH, $-NH_2$, etc.), and O_3 can easily undergo electrophilic reaction with them to generate ortho-position and para-position light-based compounds. These compounds are further oxidized to generate quinoid compounds. On the other hand, O_3 pair with electron-containing groups (such as -COOH, -NO2, etc.) aromatic compounds to attacks the meta position of these aromatic compounds. The nucleophilic reaction means that at the electron-deficient potential, the negatively charged oxygen atoms in the O_3 molecule will attack the carbon atoms containing the electron-acquiring group, and then the reaction will occur.

(2) Indirect reaction

The indirect reaction of ozone is an oxidation reaction involving free radicals [35]. During the process, ·OH is produced, and the oxidation-reduction potential is as high as 2.8eV. As a secondary oxidant, free radicals rapidly oxidize organic substances, which is a non-selective transient reaction, the reaction rate constant was observed to be 108-1010 $M^{-1}s^{-1}$, the oxidation efficiency is much higher than the direct reaction. There are three main reactions between ·OH and organic matter: electron transfer, dehydrogenation, and electrophilic addition. In the electron transfer reaction, ·OH obtains electrons from the organic matter, and it is converted into OH- itself. In the dehydrogenation reaction, ·OH seizes H from different substituents of the organic matter and transforms itself into H_2O, and the organic matter becomes an organic radical. In the electrical addition reaction, free radicals are added to the double bonds of olefins or aromatic hydrocarbons to form C-center free radicals with hydroxyl radicals on the carbon atoms (at the alpha carbon position). Studies have shown that under acidic conditions, direct oxidation is dominant. Meanwhile, under alkaline conditions, indirect oxidation is dominant, and under neutral conditions, both effects are more important.

3.2 Progress in application of oxidation with ozone

The effect of ozone on the removal of chromaticity of dye wastewater, printing, and dyeing wastewater and papermaking wastewater have been applied in recent years. Ozone can degrade chromospheres macromolecules into small molecules and finally removes them effectively. The biggest feature of dye wastewater is BOD, COD, high chroma, and odor. The methods of ozone wastewater treatment include a single ozone method, homogeneous catalytic ozone method, and heterogeneous catalytic ozone method. Liu et al. [36] used a separate ozone oxidation method and heterogeneous catalytic ozone oxidation method to treat dye wastewater. When ozone generation amount is 10 g/h, contact time is 30 min, a certain amount of MnO/zeolite, CuO/zeolite, and Fe_2O_3 are added respectively/Zeolite, the research results showed that the heterogeneous ozone catalytic oxidation treatment is superior to ozone oxidation alone, in which the removal rate of COD and chroma with MnO/Zeolite is 76.56% and 96.87%, respectively and the removal rate of chroma is significantly improved.

Ozone catalytic oxidation not only has the function of removing macromolecular substances but also has the function of improving the biochemical properties of wastewater. Hu et al., [37] used ozone oxidation to treat cellulose ethanol wastewater and improve its biochemical properties. The results showed that at an initial pH of 8 to 10, an ozone dosage of 5 g/h, a reaction time of 80 min, and a temperature of 30°C, the removal of COD and ammonia nitrogen reached 35% and 40%, respectively. The BOD_5/COD of the effluent

Materials Research Forum LLC
https://doi.org/10.21741/9781644901397-1

was increased from 0.1 to about 0.3, which greatly improved the biodegradability of the effluent. To improve the ozone utilization rate, the contact area with wastewater is increased, and the treatment efficiency is improved. The researchers proposed continuous oxidation of wastewater by ozone. Meng et al., [38] treated coking wastewater by continuously introducing wastewater and ozone. Without changing the pH of the wastewater, when the ozone dosage was 12.15 mg/L and the wastewater flow rate was 2 ml/min, COD was removed and the rate is 54.50%, COD effluent reaches <Coking Chemical Industry Pollutant Discharge Standard> (GB1617-2012).

Although ozone should be used in wastewater treatment, there are still some limitations, such as the high cost of ozone, i.e. the power consumption per kilogram of ozone is about 20-30 degrees. Moreover, the low utilization rate makes the cost of ozone treatment high [39]. The reaction selectivity between ozone and organic matter is strong, it is impossible for ozone to completely mineralize pollutants at low doses and in a short time, and the intermediate products generated by decomposition will prevent further oxidation of ozone [40]. Therefore, to strengthen the research on gas-water contact mode and contact equipment, to improve the ozone utilization rate and oxidation capacity, it has become a hot topic for the research to work for the advanced oxidation method. The AOPs of ozone is the combination of ozone oxidation and various water treatment technologies to form hydroxyl radicals (the oxidation-reduction potential of which is 2.8 eV) with higher oxidizability and lower reaction selectivity. In recent years, AOPs [41] (combined UV, H_2O_2, Fe^{3+}, homogeneous catalysts, heterogeneous catalysts, etc.) with O_3 as the main oxidant have been rapidly developed. With the development of the application of ozone technology, the research and application of ozone technology have formed an independent industry in the world, and its application prospect will be very broad.

4. Ultrasonic oxidation technology

4.1 Ultrasonic cavitation effect

Ultrasound refers to sound waves with a frequency higher than 20 kHz. Usually, the frequency in the range of 20~100 kHz is called low-frequency ultrasound, and the frequency in the range of 100 kHz~10 MHz is called high-frequency ultrasound. Ultrasonic cavitation originated in 1895 when Thormycroft and Barnaby published a report on cavitation [42]. Then the researchers began to study the effect of ultrasonic cavitation. In 1927, Loomis first proposed the effect of ultrasound to accelerate the reaction rate in chemistry and biology [43]. In 1934, researchers discovered that ultrasound could increase the rate of electrolyzed water. In 1944, Harvery et al. introduced the concept of corrected diffusion where the growth of micro bubbles was found due to unequal mass transfer across the interface during bubble vibration. In 1964, Flynn proposed the concepts of transient cavitation and steady-state cavitation [44].

Ultrasonic cavitation is an extremely complicated physical and chemical phenomenon. Ultrasonic waves consist of a series of dense and dense longitudinal waves. When ultrasonic waves pass through an aqueous solution, they alternately compress and expand the water molecules [45]. The micro-bubbles (cavitation nuclei) are existing in the liquid vibrate under the action of the sound field. When the sound pressure amplitude of the negative half-cycle of the sound wave exceeds the internal static pressure of the liquid and the micro-gas nuclei in the liquid expands rapidly and comes in succession. In the positive phase pressure of the sound wave, the gas core is compressed by adiabatic instantaneously until it collapses (also known as annihilation) [46]. At the moment of collapse, hot spots appear in the gas core and the tiny space around it, forming a high temperature and high-pressure area. A series of dynamic processes occurring under the action of the sound field is called ultrasonic cavitation [47]. The intensity of the cavitation effect is related to the temperature, pressure, cavitation core radius, gas content, viscosity, ultrasonic sound intensity, and frequency of the medium.

Even at normal temperatures, cavitation bubbles can generate temperatures up to 5000 K and pressures of 50 MPa. Generally, the bubble annihilation time is very short, only a few microseconds, the temperature change rate reaches 109 K/s. It is accompanied by a powerful shock wave and an instant glow discharge process in the liquid. According to the process of ultrasonic cavitation, the process of ultrasonic cavitation can be divided into transient cavitation and steady-state cavitation. Transient cavitation can only occur under the effect of relatively large sound intensity (greater than 10 W/cm^2), and can only exist for one or a few sound wave periods. The bubble expands sharply during the expansion phase of the ultrasound field and shrinks rapidly during the compression phase. Both expansion and compression can be regarded as adiabatic processes. The accompanying high temperature and high pressure when transient cavitation occurs can provide a physical basis for sono radicals, supercritical water, and son luminescence. Steady-state cavitation mainly refers to the dynamic behavior of cavitation bubbles containing gas and steam. Oscillation under the action of sound waves can continue for multiple sound wave periods. The chemical effect at relatively low sound intensity is believed to be due to steady-state cavitation.

Materials Research Forum LLC
https://doi.org/10.21741/9781644901397-1

4.2 Ultrasonic oxidation mechanism

4.2.1 Ultrasound reaction zone

The degradation of organic pollutants by ultrasonic oxidation is not a direct effect of sound waves but is closely related to the collapse of cavitation bubbles generated by liquids [48]. Cavitation bubbles are generated using sound wave radiation solutions with frequencies between 15 kHz and 1 MHz [49]. The long-lived cavitation bubble can be divided into three areas: cavitation bubble, gas-liquid interface surrounding the cavitation bubble, and bulk liquid phase. The schematic diagram of the mechanism of ultrasonic oxidation is shown in Fig. 3.

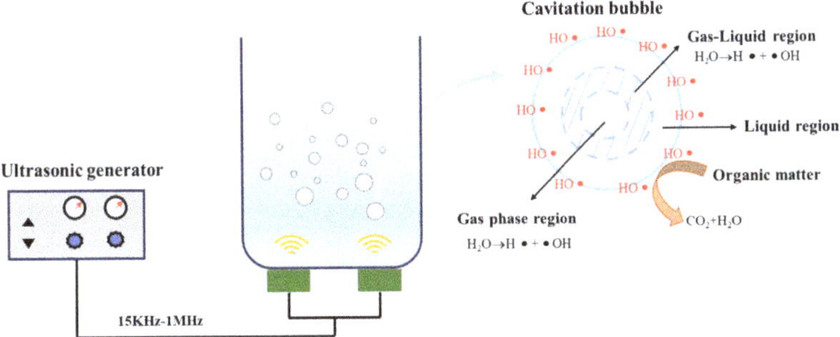

Fig.3 Schematic diagram of the mechanism of ultrasonic oxidation.

Cavitation bubble core contains water vapor, dissolved gas, and volatile solute in this area. In addition to the high temperature of 5000 K and the high pressure of 50 MPa generated when the cavitation bubble collapses, there are also ·OH and ·H radicals generated by the pyrolysis of water vapor. The high temperature and pressure and the strong oxidizing atmosphere formed by free radicals cause pyrolysis and free radical reactions in this area [50]. This creates an extreme physical environment for the degradation of organic matter, which can greatly promote the redox reaction so that some reactions that require higher temperatures and pressures can proceed smoothly under normal conditions.

At the gas-liquid interface: this area is the transition area between the gas phase of high-temperature and high-pressure cavitation bubbles and the bulk solution at room temperature and pressure. When the cavitation bubble collapses, the temperature in this area is above 2000 K. Both pyrolysis and free radical reactions will occur [51]. If the concentration of the solute is high, the pyrolysis reaction is dominant, and on the contrary, the free radical reaction is dominant. At the same time, there may be transient supercritical water, and supercritical water oxidation reaction occurs.

Liquid phase zone: the bulk solution zone at normal temperature and pressure. It is estimated that about 10% of the free radicals generated when the cavitation bubble collapses escape into this area and react with organic matter in the solution [52]. The reactions occurring in the reaction zone are also related to the nature of the reaction system. Generally, non-volatile or non-volatile hydrophilic organic compounds react with free radicals diffused from cavitation bubbles to degrade. This degradation reaction mainly occurs in the gas-liquid interface layer or bulk liquid phase; the decomposed form degrades in cavitation bubbles, but if the reactants are not volatile enough to enter the interface layer, ultrasonic irradiation cannot play a significant strengthening effect.

In addition to the formation of local high-temperature and high-pressure regions in the solution, ultrasonic cavitation also produces local high-concentration oxidizing substances, such as ·OH and H_2O_2 and supercritical water. Therefore, ultrasonic cavitation degradation of organic matter mainly includes three mechanisms: free radical oxidation, high-temperature pyrolysis, and supercritical water oxidation.

4.2.2 Free radical oxidation

In early 1982, Milino et al. verified that the formation of hydrogen radicals and hydroxyl radicals in water ultrasonic cracking using spin trapping and electron spin resonance spectroscopy. Under the action of ultrasound, at the instant when the cavitation bubble collapses, although the time does not exceed 100 ns, the high temperature, and pressure generated during the adiabatic collapse of the cavitation bubble are enough to crack the water molecules around the hot spot into ·OH and ·H radicals [53].

$$H_2O+US\rightarrow \cdot OH+\cdot H \quad (4\text{-}1)$$

·OH radicals have a higher redox potential than other common oxidants, and therefore having a higher oxidation capacity. The redox potentials of commonly used oxidants are shown in Table 1. The extremely strong oxidizing ability of ·OH radicals allows organic substances to be rapidly oxidized and degraded. At the same time, ·OH radicals can be easily recombined at the interface around cavitation bubbles, or react with volatile organic compounds in the gas phase, or soluble in the interface area of bubbles the solute reacts to form the final product, thereby achieving the purpose of degrading organic matter.

Materials Research Forum LLC
https://doi.org/10.21741/9781644901397-1

Table. 1 Oxidation-reduction potential of common oxidants.

Types of oxidants	Redox potential (φ_0/eV)
F_2	2.87
·OH	2.8
SO_4^{-}·	2.51
$S_2O_8^{2-}$	2.08
O_3	2.08
H_2O_2	1.77
HSO_5^-	1.75
MnO_2	1.68
MnO_4^-	1.68
$HClO_4$	1.63
ClO_2	1.50
CrO_7^{2-}	1.33
O_2	1.23

4.2.3 Pyrolysis

The ultrasonic chemical reaction is to accumulate the sound field energy in a tiny space through the effect of ultrasonic cavitation, generating unusually high temperature and high pressure, forming a so-called "hot spot". The high temperature and high pressure around the hot spot and the associated mechanical shear force can produce effects similar to heating and pressurization in chemical reactions to increase molecular activity [52], thereby accelerating the speed of chemical reactions. The high temperature and high pressure generated by cavitation are enough to open and break the chemical bonds (377 ~ 418 kJ/mol) of the organic matter entering the cavitation bubble, and thermal decomposition reaction similar to water phase combustion occurs in the cavitation bubble. Provide a new path for chemical reactions that are difficult to achieve under normal conditions.

4.2.4 Other auxiliary effects

It is generally believed that cavitation theory and free radical theory are the two main mechanisms of action of ultrasonic degradation of organic matter, but at the same time, ultrasonic will also produce other non-negligible effects when processing organic matter in water, including:

Mechanical shearing effect: Ultrasound is a form of mechanical energy propagation that can produce linear alternating vibration. When the cavitation bubble collapses, the medium particles will generate a large instantaneous velocity and acceleration, causing violent vibration. This severe vibration will show a strong hydrodynamic shear force macroscopically, which will make the main chain of large molecules carbon bond broken,

Materials Research Forum LLC
https://doi.org/10.21741/9781644901397-1

and at the same time, the high pressure will produce a powerful shock wave or high-speed jet in the liquid to cause the free radical to oxidize with the organic matter [54].

Supercritical water effect: the ultrasonic cavitation effect can produce a temperature of up to 5000 K and a pressure of 50 MPa. At this time, the temperature and pressure have exceeded the critical point of water (T = 647.3K, P =21.05 MPa), so at the moment when the cavitation bubble collapses, the water molecules on the surface of the cavitation bubble exceed the critical point and are in a supercritical state. Because supercritical water has special properties that are different from water in the normal state, pollutants are easily oxidized and decomposed in the supercritical state.

Thermal effect: Ultrasonic wave propagates in the medium, and its vibration energy is continuously absorbed by the medium and converted into thermal energy, which causes its temperature to rise [55]. During the ultrasonic degradation reaction, if the temperature of the reaction system is not controlled, the temperature increase of the reaction medium can be detected. When the ultrasonic energy is absorbed, it causes overall heating in the medium, local heating outside the boundary, and cavitation to form a shock wave. The local heating effect at the wavefront contributes to the activation energy of the degradation reaction, thereby helping the degradation reaction of the reactants.

Flocculation: Ultrasonic waves can promote coagulation because when the ultrasonic waves pass through the liquid medium with tiny floc particles, the suspended particles begin to vibrate with the medium, but the particles of different sizes have different vibration speeds, the particles will collide with each other, bond, increase in volume, and finally settle down. Due to the effect of ultrasound, the flocculation effect is better than the general one, and the removal rate of pollutants is higher. While the ultrasonic waves produce the above-mentioned functions, the shock waves generated willfully stir and mix the entire solution. In the process of ultrasonic degradation of organic pollutants in water, mainly through the synergy of the above processes, the pollutants are decomposed.

4.3 Application progress of ultrasonic oxidation

Although the ultrasonic degradation wastewater in the laboratory has achieved satisfactory results in technology, there are still deficiencies such as relatively high treatment cost and low energy utilization rate [56]. To further improve the degradation efficiency, reduce costs, and the combination of ultrasound and other technologies has become its development direction, thus opening up the industrial application prospects of ultrasound. Ultrasound can be used in conjunction with other technologies to produce synergistic effects and complement each other's advantages. It can be used for organic matter treatment to obtain a higher reaction rate and salinity. There are many studies on ultrasound-H_2O_2, ultrasound-Fenton, ultrasound-photo (catalytic) oxidation, ultrasound-ozone, ultrasound-electrochemistry, ultrasound-biological treatment, etc.

(1) Ultrasound-H_2O_2

When using ultrasound alone to degrade organic matter, the degradation efficiency is low, the degradation rate is slow and the energy consumption is relatively large. To improve the degradation efficiency of pollutants, some researchers added H_2O_2 to the ultrasonic reaction system and achieved satisfactory results. Zhu et al., [57] used ultrasound combined with heterogeneous catalytic oxygen to treat phenol wastewater. The results showed that the ultrasonic/H_2O_2/CuO system has an ultrasonic intensity of 4.97 W/cm^2, a dosage of H_2O_2 of 300 mg/L, and a dosage of catalyst. It was 1g/L, ultrasonic treatment at 30 °C for 240 min, and the phenol removal rate reached 89.50%. Within a certain amount of hydrogen peroxide, the degradation efficiency increases with increasing the amount of H_2O_2. When the amount of H_2O_2 is too much, the degradation efficiency decreases instead. This is because H_2O_2 can not only generate strong oxidizing free radicals, but also is a free radical consumption agent, so there is an optimized value for the amount of H_2O_2. An appropriate amount of H_2O_2 can stimulate the formation of ·OH under the action of ultrasound so that the concentration of ·OH increases, which is conducive to the improvement of degradation efficiency; and excess H_2O_2 will react with ·OH to form HOO· with oxidation ability far inferior to that of ·OH. Therefore, it is not conducive to the degradation reaction.

(2) Ultrasound-Fenton

In recent years, Fenton reagent has been applied to the treatment of dye wastewater as an AOP. However, the hydrogen peroxide utilization rate in the Fenton reaction is not high, and Fe^{2+} is easily oxidized to Fe^{3+}, which has a certain effect on the effluent color. The combined use of ultrasound and Fenton reagent can further enhance the effect of ultrasound degradation. This is due to the phenomenon of cavitation bubbles generated inside the solution when the ultrasound collapses accompanied by strong shock waves and micro jets, causing local turbulence and promoting the solution. Internal mass transfer, so that ·OH has more opportunities with organic matter is only conducive to reducing the amount of Fenton reagents used, but also can improve the processing efficiency and utilization of Fenton reagents. Ren [58] studied ultrasonic-Fenton AOP to treat beer wastewater. Under the optimal conditions of ultrasonic power of 200W, ultrasonic frequency of 45KHz, initial pH of 2, H_2O_2, and $FeSO_4$ concentrations of 70mmol/L and 7mmol/L, respectively, After 20 minutes of ultrasonic reaction, the COD removal rate of wastewater reached up to 89.8%. Ren et al., [59] used ultrasound combined with the Fenton process to treat washing wastewater. When the ultrasonic power and frequency were 140W and 45 KHz, the concentrations of H_2O_2 and $FeSO_4$ were 1mg/L and 3mg/L respectively, the ultrasonic reaction was 1h, and the COD removal rate reached 93%.

(3) Ultrasound-ozone

Ultrasound-ozone combined technology is widely studied. The key to ozone as a strong oxidant for water treatment is that ozone can be well dispersed and dissolved in water. The introduction of ultrasound can promote the full dispersion and dissolution of ozone in the solution, improve its mass transfer efficiency, and promote the rapid decomposition of ozone, generating a large number of strong oxidizing free radicals, and the dyes will degrade under the oxidation of these free radicals. Reduce the amount of ozone added. Liu et al., [60] evaluated the effects of different operating parameters on ammonia oxidation in $SrO\text{-}Al_2O_3/O_3$ and $SrO\text{-}Al_2O_3/US/O_3$ systems. It was found that the ammonia conversion rate in both systems increased with increasing pH. Due to the synergy between $SrO\text{-}Al_2O_3$, US, and O_3, the conversion rate of NH_4^+ was highest in the $SrO\text{-}Al_2O_3/US/O_3$ system, and $SrO\text{-}Al_2O_3$/Compared with the $SrO\text{-}Al_2O_3/O_3$ system, the US/O_3 system has shorter reaction time, and lower catalyst consumption and ozone consumption. Under the best conditions, the ammonia conversion rate and gaseous nitrogen yield reached 83.2% and 51.8%, respectively.

(4) Ultrasound-electrochemistry

At present, the mechanism of ultrasonic electrochemical degradation of organic matter is not very clear, but some studies believe that under the synergy of ultrasonic and electrochemical, on the one hand, ultrasound can increase the liquid phase mass transfer coefficient and reduce the concentration polarization. On the other hand, ultrasound can not only enhance the degassing effect of the electrode surface, but also remove or improve the passivation phenomenon of the electrode, make the electrode surface continuously update or expose the reaction center, and change the electrochemical characteristics of the electrode surface. In addition, the ultrasonic cavitation effect may cause some of the chemical substances in the electrochemical reaction system to directly pyrolyze in the cavitation bubble or react with the active radicals generated in the cavitation process. Therefore, in the ultrasonic-electrochemical reaction system, it is the result of the combined effects of electrode oxidation, electro-hydrogen radical oxidation, ultrasonic cavitation, and electro-flocculation. Zhou et al. [61], investigated the catalytic effects of four types of homogeneous catalysts of NaCl, $AlCl_3$, Na_2SiO_3, and $CuSO_4$ on the ultrasonic/iron-carbon micro electrolysis synergistic system for the degradation of nitrobenzene. The results showed that $CuSO_4$ had the best catalytic effect when the initial pH of the system reaction liquid is 3.0, the mixed filler dosage is 5.0 g/L (iron-carbon mass ratio is 1:1), and the

Materials Research Forum LLC
https://doi.org/10.21741/9781644901397-1

$CuSO_4$ dosage is 800 mg/L, the degradation rate of p-nitrobenzene in the system reaching 99.2%, the treatment effect is very obvious, but the removal effect of COD is not obvious.

(5) Ultrasound-biological

At present, biochemical methods are still commonly used for the treatment of refractory industrial wastewater. However, due to the complex composition of wastewater, it usually contains some raw materials and intermediates that are toxic to microorganisms, which is easy to cause microbial poisoning. Some industrial wastewater has poor biochemical properties, so before biochemical treatment, chemical or physical and chemical methods are often needed for pretreatment. If the chemical treatment method needs to add chemical agents, the cost is higher, and the treated wastewater needs to be treated by biological methods. The adsorption method is used for pretreatment, the regeneration cost of the adsorbent is high, and if it is not properly treated after adsorption cause secondary pollution. In recent years, foreign laboratories and research institutes have tried to use ultrasonic degradation as a pretreatment of biochemical methods. For some wastewater that is difficult to be biodegraded, it can be treated by ultrasound to improve its biodegradability and then treated by conventional biochemical methods. In this way, it not only solves the problem of the high cost of using ultrasound alone but also solves the problem that the biochemical method is difficult to solve. It is complementary and has good industrial prospects.

In summary, the ultrasonic oxidation technology has the advantages of mild degradation conditions, high efficiency, wide application range, no secondary pollution, simple operation, etc. It is a potential and environmentally friendly water treatment technology and has good performance in the field of wastewater treatment. Although research on ultrasonic degradation of wastewater has made some progress, the current research is still in the basic research stage, but the technology still has problems such as low degradation efficiency, long reaction time, and high energy consumption [62], especially for the polarity, hydrophilic, and volatile pollutants. In addition, many problems need to be solved, such as the scale-up design of the reactor, enhancing the ultrasonic cavitation effect, and increasing the yield of ·OH radicals [63]. In the future development of ultrasound, the researcher should pay attention to the mechanism and kinetics of chemical ultrasound and ultrasound combined technology to degrade wastewater, and establish a mathematical model to provide a theoretical basis for its application and provide theoretical parameters for the optimal design of the reactor and the optimal selection of test conditions. To seek an effective way to improve processing efficiency.

Materials Research Forum LLC
https://doi.org/10.21741/9781644901397-1

5. Wet oxidation technology

5.1 Mechanism of wet oxidation

The wet oxidation process (WAO) was originally researched and proposed by the United States in 1958 and used to treat papermaking black liquor. Before the 1970s, to ease the reaction conditions, based on the traditional WAO, adding a suitable catalyst to the reaction can reduce the temperature and pressure required for the reaction and improve the treatment effect [16]. WAO method is mainly used for municipal sludge treatment, alkali recovery in papermaking black liquor, activated carbon regeneration, etc. After entering the 1970s, the WAO waited until the rapid development, and its application range was further expanded from the recovery of useful chemicals and energy to the treatment of toxic and hazardous wastes, especially in the treatment of toxic and hazardous substances such as phenol, phosphorus, and cyanide. Domestic research on WAO has only been conducted since the 1980s, and experimental studies have been carried out on papermaking black liquor, sulfur-containing wastewater, phenol-containing wastewater, coal-to-gas wastewater, pesticide wastewater, and printing and dyeing wastewater [64]. At present, the WAO is still in the experimental stage in China, and the wet oxidation process flow is shown in Figure 5.

Compared with conventional processing methods, WAO has a wide range of applications. It can almost effectively oxidize various types of high-concentration organic wastewater, especially wastewater that is highly toxic and difficult to degrade with conventional methods. Under appropriate temperature and pressure conditions, COD treatment efficiency can reach more than 90%. The oxidation rate is fast. The reaction residence time required for most WAO to treat wastewater is within 30 to 60 minutes. Compared with biological treatment, the residence time of wastewater in the reactor is much shorter. Therefore, the WAO processing device is relatively small, occupies a small area, has a compact structure, and is easy to manage with less secondary pollution. When WAO oxidizes organic pollutants, carbon is oxidized to CO_2, nitrogen is converted to NH_3, NO_3^-, N_2, halogen, and sulfur are oxidized to corresponding inorganic halides and sulfides, and there is no NO_X, SO_2, HCl, CO such harmful substances are generated, so the secondary pollution generated is small. The WAO can not only oxidize and decompose the organic matter in the wastewater but also achieve the purposes of deodorization, decolorization, and sterilization. The WAO process requires less energy and can recover the energy generated in the reaction. For example, the energy required to treat the organic matter in WAO is the difference between the enthalpy of the incoming and outgoing water. The heat of reaction of the system can be used to heat the feed, while the heat discharged from the

 Materials Research Forum LLC
https://doi.org/10.21741/9781644901397-1

system can be used to generate steam or heat the water. Gas is used to expand turbines to produce mechanical or electrical energy.

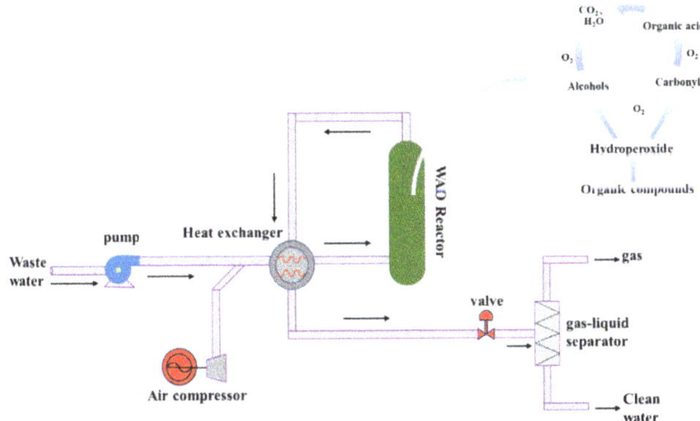

Fig.5 Wet oxidation process (WAO) flow chart.

WAO is under high temperature (125~320°C) high pressure (0.5~20 MPa) operating conditions, using oxygen or H_2O_2 in the air as an oxidant, oxidizing water in a dissolved or suspended organic matter or reduced state of inorganic substances, which are generally oxidized by air, includes two parts: the mass transfer process of oxygen in the air from the gas phase to the liquid phase; the chemical reaction between dissolved oxygen and the matrix [65], and the final products after oxidation are carbon dioxide and water. Studies generally believe that the wet oxidation reaction is a free radical reaction, and the reaction can be divided into three stages: chain initiation, chain transfer, and chain termination [66]. In the chain initiation stage, the molecular oxygen reacts with the reactant molecules to generate hydrocarbon radicals (R·); in the chain transfer stage, the radicals interact with the reactant molecules to produce ester radicals (ROO·) and hydroxyl radicals (·OH) and hydrocarbyl radicals (R·). Hydroxyl radicals have strong oxidizing properties and undergo oxidation reactions when they come into contact with organic matter; during the chain termination phase, free radicals collide with each other to form stable molecules, which interrupt the chain growth process, then it stopped. In addition to the above scholars thinking that the WAO reaction is a chain reaction, some scholars believe that according to the changes of free radicals in WAO, the reaction can be divided into four stages: induction period, proliferation period, degradation period, and end period. The whole reaction process is shown in the following reaction formula (see equations 5-1 to 5-9).

Induction period:

$$RH+O_2 \rightarrow R\cdot+HOO\cdot \text{(RH is organic matter)} \tag{5-1}$$

$$RH+O_2 \rightarrow 2R\cdot+H_2O_2 \tag{5-2}$$

Proliferative phase:

$$R\cdot+O_2 \rightarrow ROO\cdot \tag{5-3}$$

$$ROO\cdot+ RH \rightarrow ROOH+ R\cdot \tag{5-4}$$

Degeneration:

$$ROOH \rightarrow RO\cdot+\cdot OH \tag{5-5}$$

$$2ROOH \rightarrow R\cdot+ RO\cdot+H_2O \tag{5-6}$$

End period:

$$R\cdot+ R\cdot \rightarrow R\text{-}R \tag{5-7}$$

$$ROO\cdot+ R\cdot \rightarrow ROOR \tag{5-8}$$

$$ROO\cdot+ ROO\cdot \rightarrow ROH+ R_1COR_2+ O_2 \tag{5-9}$$

In addition, some research scholars believe that the wet oxidation of organic matter first generates ·OH radicals, ·OH radicals react with water to generate ·OH radicals, and then react with organic matter to generate small molecule acids, which are further oxidized to CO_2 and H_2O.

At present, the research focus of wet oxidation mainly focuses on the catalytic oxidation method (CWAO), and the key lies in the research and development of catalysts. It is generally believed that in the treatment of wastewater, the conventional WAO doped with a catalyst can better improve the treatment efficiency. CWAO catalysts are generally divided into three categories: metal salts, metals, oxides, and composite oxides, which can be divided into homogeneous CWAO and heterogeneous CWAO according to the form in which they exist in the system. The most researched CWAO catalysts in the early period were homogeneous. Because the catalyst and wastewater are in the same phase in a homogeneous catalytic reaction, the reaction temperature is milder, the reaction performance is more specific, and there is a specific selectivity. Zeng et al., [67] used a WAO to pretreat high-concentration pharmaceutical wastewater. The results showed that the addition of a homogeneous catalyst in the test can greatly improve the removal rate of COD, and the final removal rate reached 76.50%. Homogeneous catalysts have the advantages of high activity and fast reaction speed, but due to the introduction of metal ions into the system, causing secondary pollution, they usually require subsequent treatment to recover the catalyst.

Heterogeneous catalysts exist in solid form. Compared with homogeneous catalysts, heterogeneous catalysts not only have high activity and high stability but also have the characteristics of easy separation. Therefore, in recent years, the research of heterogeneous catalysts has received widespread attention. Wang et al., [68] studied the removal of bimetallic-wet oxidation of inorganic nitrogen in the concentrated filtrate of landfill leachate. The research results showed that, under suitable conditions, this method can convert 97.10% of NO_3^--N in the system into NO_2^--N. The removal rates of NO_2^--N and NH_4^+-N reached 99.03% and 69.23%, respectively. In the actual concentration treatment, the removal rate of inorganic nitrogen reached more than 72%, and the treatment effect was obvious. Wu [69] used catalytic wet oxidation to treat simulated space station condensate wastewater and sanitary wastewater. The reaction temperature was 120 °C, the pressure was 0.6MPa, the catalyst dosage was 10g, and the stirring speed was 200rpm. The TOC removal rate reached 96.01%; when the pressure was 1.2 MPa, without changing other conditions, the TOC removal rate of sanitary wastewater reached 98.68%.

5.2 Development trend of wet oxidation

Although the WAO after adding the catalyst has the advantages of fast reaction rate, strong degradation ability, and stable treatment effect [70]. It is also accompanied by many defects. Common deficiencies include high requirements for operating costs, high investment costs brought about by the development of catalytic materials, and possible production of toxic intermediate products. In the development of catalysis, an important issue is how to effectively maintain the activity of the catalyst for a long time [71] to avoid catalyst deactivation. During the long-term use of the catalyst, its activity will decrease. The cause of industrial catalyst deactivation can be summarized as the following three conditions: catalyst poisoning, carbon deposition, or coking, the physical structure changes. Catalyst poisoning produces certain components in the reaction system can be firmly and reversibly or irreversibly adsorbed on the active center on the surface of the catalyst, and the catalyst will be temporarily or permanently poisoned. The temporarily poisoned catalyst can be regenerated to restore activity; carbon deposition or coking is manifested as non-volatile substances generated during the reaction, such as high polymer or carbonization, cover the active center or block the pores of the catalyst; the physical structure changes mainly include catalyst grain growth, reduced specific surface area, and surface fusion [72]. In addition, there are relatively few studies on the reaction mechanism and kinetics in the wet oxidation process, and these aspects should be strengthened in the future. CWAO has a wide range of application backgrounds. CWAO catalysts are developing towards multi-component, high activity, low cost, and good stability.

6. Supercritical water oxidation technology

6.1 Supercritical water and its characteristics

Under normal circumstances, the common states of water are divided into three types: liquid water, steam, and ice, and exist in one of them. However, when the temperature and pressure of the water exceed the critical point (T=374.5 °C, P=22.05MP) [73], the water in this state is called supercritical water. The density of water at normal temperature and pressure is almost unchanged, and it is considered to be a polar solvent. Most electrolytes can be dissolved in it, but the gas and most non-polar organic substances are slightly soluble or insoluble. In the supercritical state, these series of properties have undergone tremendous changes.

Near the critical point, the density of water changes significantly with temperature. The density of supercritical water (SCW) is between the density of liquid water (1 g/cm^3) and the density of low-pressure water vapor (0.0011 g/cm^3). The critical point density is 0.326 g/cm^3. As the temperature increases, the dielectric constant of water decreases. Compared with the normal water, SCW has solubility close to organic solvents [74]. near the critical point, small changes in temperature and pressure will make the viscosity of SCW, a huge change has occurred, and the viscosity of SCW is close to that of gas, making SCW have high fluidity and diffusion coefficient and mass transfer properties approaching high-temperature gas [75]. The ion product also changes significantly with changes in water temperature and density. Taking the ion product at 25 MPa as an example, when the temperature range is from 250°C to 360°C, the ion product of water rises to about 11 and appears acidic, but when the water is in a supercritical state, the ion product is a little lower than that of normal water, about ten orders of magnitude so free radical reactions dominate. The macroscopic properties of water are determined by the microstructure of water. Intermolecular hydrogen bonding is one of the important factors that affect many special properties of water. Under supercritical conditions, the hydrogen bonding in water is weakened, and water molecules exist in the form of clusters. The increase in temperature and the decrease in water density reduces the number of molecules in the water molecule cluster. The weakening of hydrogen bonds reduces the intermolecular force and molecular dipole distance, and the structure of water molecules tends to be closer to nonpolar molecules [76]. These properties make SCW an excellent solvent for a series of chemical reactions.

6.2 Mechanism of oxidative degradation of organic matter in supercritical water

The SCW oxidation method was first proposed by Modell of the Massachusetts Institute of Technology and Gloyna of the University of Texas in the 1980s [77]. Gloyna and ECO WASTE Company cooperated to develop the supercritical water oxidation process in 1990 and began to process industrial wastewater from 1994. With the deepening of supercritical water oxidation technology, the supercritical water oxidation process is gradually applied in the treatment of organic wastewater. The typical supercritical water oxidation process flow is shown in Figure 6.

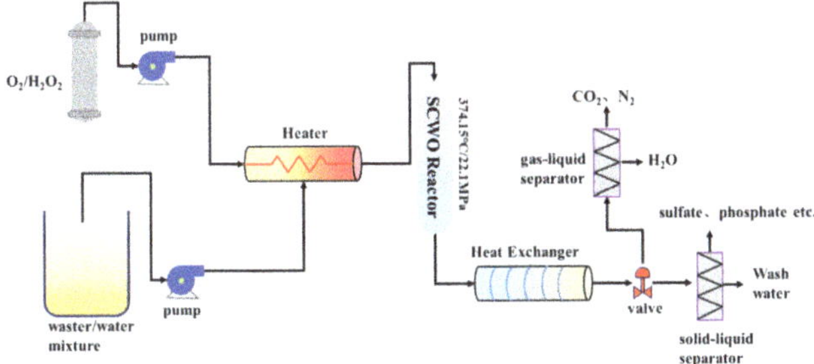

Fig.6 Schematic diagram of the supercritical water oxidation process.

The supercritical water oxidation method (SCWO) uses water in a supercritical state (temperature is higher than 374°C, the pressure is greater than 22.1 MPa), the water density, dielectric constant, viscosity, diffusion coefficient, and other great changes occur [77]. When organic matter and oxygen can be dissolved in it, the water-gas-liquid interface disappears into a homogeneous system, and oxygen or hydrogen peroxide is used as an oxidant to cause a free radical reaction to degrade organic matter [78]. The SCWO reaction of organic matter is based on the free radical reaction mechanism. The basic reaction steps are as follows: First, oxygen attacks the weakest C-H to generate free radical $H_2O\cdot$, and the reaction equations are shown in (6-1 to 6-2); second, H_2O_2 further decomposes to form hydroxyl radicals, and the reaction equation is shown in (6-3); furthermore, the hydroxyl radicals with strong electrophilicity react with hydride to generate radicals R·, and R· reacts with oxygen to generate oxidized radicals, Oxidation is used as a free radical to obtain hydrogen atoms to form peroxides. The reaction equation is shown in (6-4 to 6-6); finally, the peroxide decomposes to produce compounds with smaller molecules, and unstable compounds are further broken down and finally converted into carbon dioxide and water.

$$RH+O_2 \rightarrow 2R\cdot+H_2O\cdot \quad (6\text{-}1)$$

Materials Research Forum LLC
https://doi.org/10.21741/9781644901397-1

$RH+H_2O\cdot \rightarrow 2R\cdot +H_2O_2$ (6-2)

$H_2O_2+M\rightarrow 2HO\cdot$(M is a homogeneous or heterogeneous interface) (6-3)

$RH+HO\cdot \rightarrow R\cdot +H_2O$ (6-4)

$R\cdot + O_2\rightarrow ROO\cdot$ (6-5)

$ROO\cdot +RH\rightarrow ROOH+ R\cdot$ (6-6)

It can be seen from the above reaction that hydroxyl radicals with high activity play a key role in the oxidation process of organic matter. The number of free radicals and the level of activity determines the reaction rate and the effect of organic matter removal. However, in actual wastewater treatment, the oxidant used and the reaction conditions are not the same, and the treatment targets are different, so the role of the free radicals generated in the above stages is not the same.

6.3 Application progress of supercritical water oxidation (SCWO)

SCWO is an emerging wastewater treatment technology, which refers to the oxidation reaction of organic matter in SCWO. It has the advantages of fast reaction rate, high pollutant removal rate, and complete recycling of treated wastewater. It is often used to treat contaminants that are difficult to degrade completely by conventional methods. Li et al., [79] studied the treatment of reactive brilliant red X-3B simulated dye wastewater by supercritical water oxidation. The results of the orthogonal test showed that when the pressure is 25 MPa, the temperature is 500 °C the best removal rate of COD and chroma reached 98.28%, 99.95%. Although using SCWO technology alone is efficient, it consumes a lot of energy. To improve this situation, some researchers have combined other advanced oxidation technologies with SCWO to reduce energy consumption. Li et al., [80] used the Fenton oxidation method and supercritical water oxidation method to treat landfill leachate, and combined the Fenton method as pretreatment method and advanced treatment method with SCWO to treat landfill leachate respectively, and explored its effect on improving leachate. The removal effect of the main pollutants COD, NH_3-N, and chroma results show that the Fenton method is used in combination with SCWO as a pretreatment method and an advanced treatment method. Both of these combined methods can significantly improve the removal rate of each pollutant. However, the Fenton method is better than the pretreatment method as a pretreatment method. The combination of the two is suitable for the Fenton oxidation method. The suitable pH is 4, n (H_2O_2): n (Fe^{2+}) = 4: 1, reaction time 2 h, supercritical water Oxidation method suitable temperature 440°C, pressure 26 MPa, peroxide amount K = 3.0, the removal rate of COD, ammonia nitrogen and color reached 95.80%, 71%, 99.5% respectively.

Materials Research Foundations **102** (2021) 1-47
https://doi.org/10.21741/9781644901397-1

The SCWO has the advantage of degrading certain wastewaters in addition to being commonly used to treat conventional wastewater with good results. The researchers found that the supercritical water oxidation method has advantages in the degradation of phenol-containing wastewater. Yan et al., [81] used SCWO to efficiently treat wastewater containing p-tert-butylcatechol. When the temperature was set at 400-550 °C and the pressure was set at 18-27MPa, the COD in the wastewater was effectively degraded, and the COD was degraded. The rate reached 99.3%. The high concentration of delocalized electrons in the molecule of phenols leads to the stability of the molecule and the difficulty of opening the aromatic ring, making phenols a typical refractory pollutant in wastewater. When the temperature and pressure reach or exceed the critical point of water, the water reaches a supercritical state, the polarity of water molecules decreases, tends to be non-polar, and the solubility of organic phenols in water will be greatly improved [82]. At this time, both organic matter and oxidant are dissolved in supercritical water, so that the oxidation reaction occurs in a homogeneous system. When H_2O_2 is used as the oxidant in the supercritical water oxidation system, ·OH radicals will be excited, and then ·OH radicals can initiate free radical oxidation of phenols. ·OH radicals excite phenol into phenol radicals, and then hydroxyl radicals combine with phenol radicals to form peroxyphenol radicals, and the rearrangement of the radicals leads to aromatic ring opening [83]. Subsequently, the hydroxyl radical further excites the chain-like free radical after ring-opening, leading to oxidation.

6.3 Development trend of supercritical water oxidation

Extensive experimental researche by predecessors have shown that SCWO has many advantages, but there are still many difficulties in practical application. Mainly manifested in the following aspects: high temperature and high-pressure reaction conditions are easy to cause material corrosion, especially in the presence of an acid, the corrosion problem is more serious, but high temperature and high pressure is an essential condition of the SCWO reaction system, it is difficult to find a reaction material can withstand the corrosion of various acid solutions [84]. The solubility of the salt at normal temperature and pressure drop. Undissolved salt components accumulate and precipitate in the reactor [85], causing the reactor to block; the SCWO reaction is carried out in a high-temperature and high-pressure environment, which requires an external power supply to heat and pressurize, and the energy cost is high.

Problems such as corrosion and clogging of SCWO reaction have been the focus of future research. Also, how to improve the efficiency of the reaction is important. The reaction rate in oxidation and catalytic supercritical water oxidation ultimately improves the removal effect. The construction of the reaction system can consider the micro-reactor to

Materials Research Forum LLC
https://doi.org/10.21741/9781644901397-1

reduce the cost. The development of special materials makes the reaction device have the characteristics of high-temperature resistance, high-pressure resistance, and corrosion resistance.

7. Other advanced oxidation technologies

7.1 Electrochemical oxidation technology

7.1.1 Electrochemical oxidation mechanism

Electrochemical oxidation technology is that the modification present on the electrode surface or solution phase can promote or inhibit the electron transfer reaction on the electrode under the action of the electric field, while the modification on the electrode surface or solution phase itself does not change. In recent years, electrochemical oxidation technology has been widely used in the treatment of various refractory organic substances. A large number of domestic and foreign researchers have conducted systematic studies on the influencing factors and degradation mechanisms of their electrocatalytic performance. Compared with other oxidation processes, the electrochemical oxidation process is relatively complicated, and the oxidation of organic matter is multi-path. According to different oxidation pathways, the electrochemical oxidation process can be roughly divided into electrochemical direct oxidation and electrochemical indirect oxidation [86].

(1) Electrochemical direct oxidation

The electrochemical direct oxidation method utilizes the high potential of the anode to oxidize and degrade organic or inorganic pollutants in wastewater. During the reaction, the pollutants directly transfer electrons with the electrode [87]. During the oxidation process, pollutants are oxidized to varying degrees. Some toxic pollutants are oxidized into non-toxic pollutants in the reaction, or non-biochemically pollutants can be oxidized into biochemically treatable substances, which is conducive to further biochemical treatment. In this way, the pollutants are converted, which is called an electrochemical conversion. Some pollutants are completely oxidized into stable inorganic substances (CO_2, H_2O), which is called complete oxidation (mineralization) or electrochemical combustion. To save costs and reduce energy consumption, general pollutants only need to be oxidized into biodegradable substances.

Studies have shown that the oxidation reaction mechanism and products of organic matter on the metal oxide anode are related to the valence state of the anode metal oxide and the type of oxide on the surface. The higher-valent metal oxide MO_{x+1} generated on the anode of the metal oxide MO_x is conducive to the selective oxidation of organic matter to form oxygen-containing compounds; the free radical $MO_x(\cdot OH)$ generated on the MO_x anode is conducive to the oxidation and combustion of organic matter to generate CO_2 [87].

The specific reaction mechanism is as follows: in the potential area of the oxygen precipitation reaction, high-valence oxides may be formed on the surface of the metal oxide, so there are two states of active oxygen on the anode, namely, adsorbed hydroxide radicals and high-valence oxidation in the crystal lattice Oxygen. The anode surface oxidation process is carried out in two stages. First, the H_2O or $\cdot OH$ in the solution is discharged on the anode and forms adsorbed hydroxyl radicals. The reaction equation is shown in (7-1); then the adsorbed radicals and the existing ones on the anode Oxygen reacts, and transfers the oxygen in the free radical to the metal oxide lattice to form MO_{x+1}, the reaction equation is shown in (7-2); when there is no organic matter in the solution, the two states Activated oxygen undergoes oxygen precipitation reaction according to equations (7-3) and (7-4); when oxidizable organic substance R exists in the solution, the reaction equations are shown in (7-5) and (7-6).

$$MO_x + H_2O \rightarrow MO_x(\cdot OH) + H^+ + e^- \quad (7\text{-}1)$$

$$MO_x(\cdot OH) \rightarrow MO_{x+1} + H^+ + e^- \quad (7\text{-}2)$$

$$MO_x(\cdot OH) \rightarrow 0.5O_2 + MO_x + H^+ + e^- \quad (7\text{-}3)$$

$$MO_{x+1} \rightarrow MO_x + 0.5O_2 \quad (7\text{-}4)$$

$$R + MO_x(\cdot OH)_y \rightarrow MO_x + CO_2 + y\ H{+} + e^- \quad (7\text{-}5)$$

$$R + MO_{x+1} \rightarrow MO_x + RO \quad (7\text{-}6)$$

To completely mineralize the pollutants, the concentration of oxygen vacancies in the oxide lattice on the anode surface must be high enough, and the concentration of adsorbed light radicals should be close to zero, according to this reaction, the reaction speed of (7-2) must be faster than that of (7-1). At this time, the current efficiency of the reaction depends on the ratio of the reaction (7-6) to the reaction (7-4). Since they are all pure chemical steps, the current efficiency of the reaction (7-6) will be independent of the anode potential but depends on the reactivity and concentration of organic matter and electrode materials. Also, the anode used for the electrochemical combustion reaction must have a high concentration of adsorbed hydroxyl radicals on the surface, and the concentration of oxygen vacancies in the oxide lattice is low [88]. The current efficiency of the reaction at this time depends on the ratio of the reaction (7-5) to the reaction (7-3). Since both reactions are electrochemical steps, the current efficiency of the reaction (7-5) depends not only on the organic matter. The nature and concentration of the electrode, as well as the electrode material, are also related to the anode potential.

(2) Electrochemical indirect oxidation

Electrochemical indirect oxidation can be divided into indirect anodic oxidation and indirect cathodic oxidation according to different oxidation paths. Indirect anodic oxidation is a strong oxidant produced by the oxidation reaction at the anode, indirectly oxidizing organic matter in water to achieve the purpose of enhanced degradation [89]. Since indirect electro-oxidation not only exerts anodizing effect to some extent but also utilizes the generated oxidant, the treatment efficiency is greatly improved. Indirect anodization is divided into two categories.

One is the direct use of anions. For example, the direct oxidation of chloride ions at the anode produces new ecological chlorine or further formation of hypochlorite, thereby causing strong oxidation and degradation of organic matter in the water. This method has been used for phenol-containing wastewater treatment and high-salt organic wastewater treatment. However, some people think that a potential disadvantage of this method is that some organic substances may be chlorinated during the degradation process, and the toxicity is increased, causing secondary pollution. The other is the use of reversible redox electricity to indirectly oxidize organic matter. Commonly used pairs are: Co^{3+}/Co^{2+}, Fe^{3+}/Fe^{2+}.

Cathode indirect oxidation is through the reduction of hydrogen peroxide or ferrous ions produced by the cathode at an appropriate electrode potential, and the addition of appropriate reagents to generate a Fenton-like reaction, thereby indirectly degrading organic matter. According to the different products produced by the cathode, it can be divided into two categories: cathode produces hydrogen peroxide; iron ion cathode electrolytic reduction produces ferrous ions. Cathode produces hydrogen peroxide. Under acidic conditions, O_2 generates electrons to generate H_2O_2. See (7-7) for the reaction formula, under alkaline conditions, O_2 and H_2O get electrons to generate HO_2^- and HO_2^- reacts further to generate H_2O_2. See (7-8) and (7-9) for reaction formulas. Under alkaline conditions, the tank pressure is very low, which can reduce energy consumption and improve current efficiency. But because the oxidation potential of hydrogen peroxide is not very high, the oxidation capacity is limited. Many researchers consider adding metal catalysts such as ferrous ions to catalyze the generation of light radicals by hydrogen peroxide to form electric Fenton reagents. See (7-10) and (7-11) for the reaction formulas. Compared with the conventional chemical Fenton reagent, this method does not need to add hydrogen peroxide, and has the following advantages: through the control of electrocatalytic conditions, it can accurately control the production of H_2O_2 and the rate of organic matter degradation, avoiding the possible harm of H_2O_2 transportation and transfer. The nascent H_2O_2 has a stronger oxidation capacity and a faster reaction speed. At present,

Materials Research Forum LLC
https://doi.org/10.21741/9781644901397-1

the current efficiency generated by hydrogen peroxide on the cathode used is between 50% and 95%, and some organic pollutants can be completely removed.

$$O_2+H^++2e^- \rightarrow H_2O_2 \quad (7\text{-}7)$$

$$O_2+H_2O+2e^- \rightarrow HO_2^-+OH^- \quad (7\text{-}8)$$

$$HO_2^-+H_2O \rightarrow H_2O_2+OH^- \quad (7\text{-}9)$$

$$M+H_2O_2+H^+ \rightarrow M^++\cdot OH+H_2O \quad (7\text{-}10)$$

$$M+H_2O_2 \rightarrow M^++\cdot OH+OH^- \quad (7\text{-}11)$$

Iron ion cathodic electrolysis mainly generates a Fenton-like reaction between ferrous ions and added hydrogen peroxide to regenerate iron ions and then reduces iron ions to ferrous ions through electrode reaction, which can continuously promote the catalytic reaction. The reaction of ·OH radical, Fe^{3+}, Fe^{2+} is degraded, and the reaction formula is as (7-12) to (7-17). This method has found application in actual industrial wastewater. The main problem is that the amount of acid is large. As the concentration of ferrous ions increases, the current efficiency drops sharply, producing more iron sludge.

$$Fe^{2+}+H_2O_2 \rightarrow Fe^{3+}+OH^-+\cdot OH \quad (7\text{-}12)$$

$$Fe^{3+}+e^- \rightarrow Fe^{2+} \quad (7\text{-}13)$$

$$RH+\cdot OH \rightarrow R_1\cdot+H_2O \quad (7\text{-}14)$$

$$RH+\cdot OH \rightarrow R_2\cdot+H_2O \quad (7\text{-}15)$$

$$R_1\cdot+Fe^{3+} \rightarrow R_1^++Fe^{2+} \quad (7\text{-}16)$$

$$R_2\cdot+Fe^{2+} \rightarrow R_2^++Fe^{3+} \quad (7\text{-}17)$$

7.1.2 Application progress of electrochemical oxidation

In the early 1940s, some people abroad proposed the use of electrochemical methods to treat wastewater, but due to lack of electricity, the cost is relatively high, and the development is relatively slow. Since the 1960s, with the development of the power industry, the electrochemical method was used in the research of wastewater treatment processes. Traditional electrochemical oxidation mainly uses Cl_2 and NaClO generated in the electrolysis process to oxidize pollutants in wastewater. Due to the limited oxidation capacity of Cl_2 and NaClO and the limited electrode life, further, development is restricted. In the 1980s, the concept of AOP was proposed, and it was pointed out that the generation of hydroxyl radicals by the reaction is a key step in the effective treatment of refractory pollutants. Scientists began to study the use of electrochemical methods to produce more oxidizing oxidants without secondary pollution, such as ·OH, H_2O_2, Fenton

Table 2. Electrochemical oxidation method to remove wastewater with different pollutants

Materials Research Forum LLC
https://doi.org/10.21741/9781644901397-1

Pollutant	Applied processes	Parameters assessed	Result	Reference
Papermaking wastewater	electrochemical oxidation(anode: RuO_2/SnO_2,cathode: stainless steel)	time 60min and current density was 5mA/cm^2	COD removal of 54.6%	[90]
Papermaking tobacco sheet wastewater	Electrocoagulation+ electrochemical oxidation(anode:β-PbO_2 , athode: stainless steel)	EC time 6min and current density was and 40mA/cm2, 2.5 g/L of NaCl	COD removal of 83.9%, BOD_5/COD ratio increased from 0.06 to 0.85	[91]
Dyeing Wastewater	Photocatalysis+electrochemical oxidation	pH=2,chloride concentration of 2 g/L, current density of 10 mA/cm^2, time 180min	COD、TOC removal of 96%、68%	[92]
Petroleum refinery wastewater	electrochemical oxidation (anode:Ru-MMO)	time 40min and current density was 5mA/cm^2	COD、phenol removal of 96.04%、99.53%	[93]
Coking wastewater	electrochemical oxidation (anode:Ti/RuO_2-IrO_2)	area-volume ratio of 14.44m^2/m^3 and electrode distance of 0.5cm	COD、NH_4^+-N almost removed	[94]
Landfill leachate	electrochemical oxidation (anode:BDD)	$Cl^-:SO_4^{2-}$ =237:1, pH=3 the current density was and 50、75、100mA/cm^2	that current density had considerable effects on COD	[95]

 Materials Research Forum LLC
https://doi.org/10.21741/9781644901397-1

reagent (H_2O_2+Fe^{2+}), and ozone (O_3). In the early 1990s, the current electrochemical oxidation process was gradually developed and formed. In recent years, the electrochemical oxidation method has been widely used in the treatment of landfill leachate, tannery wastewater, printing and dyeing wastewater, oil refining wastewater, and papermaking wastewater in Table 2.

7.1.3 Development trend of electrochemical oxidation

The research and development of the electrochemical oxidation treatment method of organic wastewater, the goal is the soil industry. This industrialization process is realized through the conversion of experimental research to applied research. Starting from solving practical problems, the basic theoretical research of the electrochemical oxidation method, electrode material development, electrochemical reactor development, and electrochemical oxidation process research is the current development trend [96]. The study of the electrochemical oxidation mechanism of organic pollutants refers to the electron transfer of organic compound molecules on the electrode surface, or the interaction of strong oxidizing species and organic molecules produced at high potential, and the types of strong oxidizing species produced in electrocatalytic systems. And methods; the development of electrode materials mainly includes the preparation and optimization of electrodes, electrocatalytic activity, and selectivity, electrode life, etc., the research of electrolytic cell structure and the development of high-efficiency electrolytic reactors are mainly based on the developed electrodes and knowing a clearer oxidation mechanism, conducting systematic studies on the rational design of electrode structures and reactors, and optimizing operating conditions; conducting application studies on specific electrochemical oxidation systems, such as studying partial electrochemical degradation and complete oxidation of actual wastewater systems In the process, systematically investigate the influence of factors such as current density, temperature, pH, electrolyte, wastewater concentration, mass transfer method and speed, residence time, etc., and design the best process route.

7.2 Persulfate oxidation technology

7.2.1 Persulfate activation

The AOP in the traditional sense is a reaction process to generate ·OH. The AOP based on $SO_4^{-}\cdot$ has been developed in recent years. Its oxidation-reduction potential is 2.8 ~3.1eV, which is equivalent to that of ·OH. It is more active under alkaline conditions. The source of $SO_4^{-}\cdot$ is mainly the breaking of the peroxy bond (O-O) of persulfate (PDS) and permonosulfate (PMS) [97].

The structure of persulfate and persulfate is very stable. To break the peroxy bond to generate $SO_4^{-}\cdot$, it needs to be activated by introducing substances. According to the different activation methods, it can be divided into thermal activation, alkali activation, metal activation, organic activation, and ultraviolet activation [98]. Thermal activation can accelerate the breaking of peroxy bond by increasing the temperature of the reaction system and promote the formation of $SO_4^{-}\cdot$. The method is fast and effective and can achieve the purpose of effectively degrading the target pollutants by controlling the temperature. Alkali activation can generate $SO_4^{-}\cdot$ under alkaline conditions, and it can also generate OH and hydrogen peroxide, but studies have found that alkali activation efficiency is low. The alkali activation mechanism is shown in formulas (7-18) to (7-20). The genus activation is similar to the Fenton reaction. Cobalt ion activates persulfate to generate $SO_4^{-}\cdot$ reaction, as shown in formula (7-21). Manganese ions, iron ions [99], silver ions, cerium ions, etc. can catalyze the formation of $SO_4^{-}\cdot$ from persulfate and permonosulfate. Its catalytic activity is related to metal solubility and concentration. Organic matter-activated phenols and quinones can be used as catalysts to activate $SO_4^{-}\cdot$ to improve the removal efficiency of target pollutants. Tetrachlorophenol activates persulfate to degrade nitrobenzene, and the removal efficiency may be improved because deprotonated tetrachlorophenol attacks persulfate and promotes the breaking of peroxy bonds. Ultraviolet activation is to obtain energy by introducing ultraviolet light, and the peroxy bond breaks to form $SO_4^{-}\cdot$. The reactions are shown in formulas (7-22) and (7-23). Compared with other activation methods, the ultraviolet activation effect is better, and will not introduce impurities. For persulfate, the quantum yield of $SO_4^{-}\cdot$ under anaerobic environment is 1.8, and for permonosulfate, the quantum yield of $SO_4^{-}\cdot$ and $\cdot OH$ is 0.52.

$$S_2O_8^{2-}+H_2O\rightarrow SO_5^{2-}+SO_4^{2-}+H^+ \text{ (Under alkaline conditions)} \quad (7\text{-}18)$$

$$SO_5^{2-}+H_2O\rightarrow H_2O^{-}+SO_4^{2-}+H^+ \text{ (Under alkaline conditions)} \quad (7\text{-}19)$$

$$S_2O_8^{2-}+H_2O^{-}\rightarrow SO_4^{-}\cdot+O_2^{-}\cdot+H^+ \quad (7\text{-}20)$$

$$Co^{2+}+HSO_5^{-}\rightarrow Co^{3+}+SO_4^{-}\cdot+OH^{-} \quad (7\text{-}21)$$

$$S_2O_8^{2-}\rightarrow 2\,SO_4^{-}\cdot(UV) \quad (7\text{-}22)$$

$$HSO_5^{-}\rightarrow SO_4^{-}\cdot+\cdot OH \quad (7\text{-}23)$$

7.2.2 Based on $SO_4^{-}\cdot$ mechanism of oxidative degradation of organic pollutants

$SO_4^{-}\cdot$ has strong oxidizing properties and can partially inorganic-organic compounds. In recent years, there has been more and more research on the application of $SO_4^{-}\cdot$ to remove organic pollutants. The degradation path of organic pollutants in the AOP based on $SO_4^{-}\cdot$ is similar to the AOP based on $\cdot OH$, mainly divided into hydrogen abstraction reaction, addition reaction, and electron transfer reaction [100]. Relative to the $\cdot OH$ oxidation pathway, the oxidation of $SO_4^{-}\cdot$ tends to occur electron transfer reactions, and $\cdot OH$ oxidation tends to occur hydrogen abstraction and addition reactions. $SO_4^{-}\cdot$ Reaction with aromatic compounds mainly through electron transfer. Studies have shown that $SO_4^{-}\cdot$ degrades alkanes, alcohols, and ethers by hydrogen abstraction reaction, and the measured reaction rate constant is 10^7~ 10^9 $M^{-1}s^{-1}$, as shown in formula (7-24). $SO^{4-}\cdot$ through addition Reaction to degrade olefins, the measured reaction rate constant is about 10^6 $M^{-1}s^{-1}$, as shown in formula (7-25). $SO_4^{-}\cdot$ Degradation of amines through electron transfer reaction, the measured reaction rate constant is about 10^7~ 10^9 $M^{-1}s^{-1}$, as shown in formula (7-26). $SO^{4-}\cdot$ can exist stably when the pH value is less than 8.5. When the pH value is greater than 8.5, it will be transformed with ·OH by hydroxide or water molecules, as shown in formula (7-27) and formula (7-28).

$$SO_4^{-}\cdot + RH \rightarrow HSO_4^{-} + R\cdot \quad (7\text{-}24)$$

$$SO_4^{-}\cdot + H_2C{=}CH_2 \rightarrow {}^{-}OSO_2OH_2C{=}CH_2\cdot \quad (7\text{-}25)$$

$$SO_4^{-}\cdot + (CH_3CH_2)_2NH \rightarrow (CH_3CH_2)_2NH^{+}\cdot + SO_4^{2-} \quad (7\text{-}26)$$

$$SO_4^{-}\cdot + OH^{-} \rightarrow SO_4^{2-} + \cdot OH \quad (7\text{-}27)$$

$$SO_4^{-}\cdot + H_2O \rightarrow HSO_4^{-} + \cdot OH \quad (7\text{-}28)$$

7.2.3 Application progress of persulfate oxidation

Persulfate activation technology is widely used in the removal of environmental pollutants due to its strong stability and oxidation, not only for aromatics, heterocycles, chloroalkanes in polluted waters of pesticides, printing, and dyeing, and leather industry. The removal of alkanes and ethers has a good effect, and it also shows a strong oxidation effect on contaminated Cd^{2+}, As^{3+}, Mn^{2+}, and other inorganic pollutants, and also shows a strong In-situ chemical repair capability. See Table 3 for some application studies.

7.2.4 Development trend of persulfate oxidation

Activated sulfate oxidation method, as an AOP that has attracted attention in the field of sewage treatment in recent years, has its obvious advantages, but at the same time, it also has deficiencies. The main problems are as follows: Although activated sulfate oxygen has the advantages of rapid reaction, strong oxidizability, and simple operation [106], it has the problems of limited reaction conditions, inability to completely degrade pollutants, a dosage of reagents, and high cost. Therefore, this method cannot be widely used in practical projects; persulfate as a strong oxidant can degrade most pollutants, but it cannot be effectively removed for some substances, and the degree of degradation of different pollutants is also different. A lot of. The degradation efficiency of persulfate is related to the composition and structure of the pollutants. Therefore, to obtain the ideal removal effect, how to select the appropriate activation method according to the characteristics of the organic pollutants has become a hot issue for many researchers. Activating the persulfate oxidation method will also produce residual SO_4^{2-} and H^+. Generally speaking, SO_4^{2-} is not harmful, but a large amount of discharge will cause corrosion when the water quality is acidic, which will cause certain harm to soil and pipelines. Therefore, full consideration and attention should be given to the subsequent treatment of residual $SO_4^{2-}\cdot$. To activate persulfate to generate more free radicals, the catalyst plays an important role. Compared with homogeneous catalysts, it is difficult to remove after the reaction, and it is prone to secondary pollution. Low chemical or energy demand, reusability, no secondary pollution, and other characteristics. Heterogeneous catalyst combined with a homogeneous catalyst to activate oxidant to remove organic pollutants is an important research object in the future.

Table 3. Persulfate oxidation method to remove wastewater with different pollutants.

Pollutant	Applied processes	Parameters assessed	Result	Reference
Groundwater of BTEX and TCP	Persulfate oxidation	mixed oxidant solution (persulfate, ferrous sulfate, and citric acid) of 6 m^3, time of 24 h	TCP、BTEX removal of 61.4%、about 100%	[101]
Azo dye wastewater	iron-activated persulfate oxidation	pH=4.8, Fe^{2+}=1.64 mMol $S_2O_8^{2-}$=84.87 mMol	Mineralization of 35.14%	[102]
Coal Chemical Industry wastewater	thermal and alkali of Persulfate oxidation	thermal and alkali of Persulfate, pH=12, time of 10 min、20 min	MOP degradation ratio of 88.4% and 93.8% within 10 min and 20 min,	[103]
cyanide-containing organic wastewater	Electrochemical Oxidation + Persulfate oxidation	persulfate concentration 0.1 Mol, current density 10 mA /cm^2, pH=5.6, temperature 40 °C , time of 24 h	COD、TOC、CN^- removal of 95.8%、87.8% 、98.4%	[104]
DDNP industrial wastewater.	microwave irradiation+ Persulfate activated by Fe0	PS =6 g/L, Fe^0 = 0.4 g/L, MW power = 450 W , pH = 3	COD、TOC removal of 69.52%、57.27%	[105]

Materials Research Forum LLC
https://doi.org/10.21741/9781644901397-1

Conclusion

It is concluded that, to achieve a sustainable water environment, advanced oxidation technologies (AOPs) are highly efficient technologies. It is observed that the basic theories for elimination or degradation of pollutants are not the same, as the actual pollution system is complex and diverse. Thus, many different APOs have been reported for the removal of pollutants. If we consider using a single system to eliminates or degrade pollutants than it could be very challenging for researchers. The combined use of AOPs or combined with other technologies, can play a synergistic role and strengthen the pollutant treatment effect. Therefore, it has high treatment costs, harsh reaction conditions, and complex reactors system to remove pollutants. Recent literature has found that the challenges over the period of simple ozonation technologies have been minimized after accepting the hybrid ozonation system. Likewise, many new technologies have been reviewed with their barrier and challenges for the removal of pollutants. In common, studies found low-cost and high-efficient oxidants and catalysts with good activity, stability, high efficiency, and wide applicability are highly prioritized to further improve the water quality. In those aspects, the oxidation reaction mechanism of different pollutants has been explored in this review paper. Finally, from the application point of view, AOP can become the most adaptable application for the wastewater treatment plants, as most of the limitations have been overcoming using hybrid AOPs. At last, this review throws light on the role of AOPs for establishing water sustainability.

References

[1] Y. Sun, S. Zhu, W. Sun, H. Zheng, Degradation of high-chemical oxygen demand concentration pesticide wastewater by 3D electrocatalytic oxidation, J. Environ. Chem. Engin. 7(2019) 103276. https://doi.org/10.1016/j.jece.2019.103276

[2] K. S. Varma, R. J. Tayade, K. J. Shah, P. A. Joshi, V. G. Gandhi, Photocatalytic degradation of pharmaceutical and pesticide compounds (PPCs) using doped TiO_2 nanomaterials: A review, Water Energy Nexus 3(2020)46-61. https://doi.org/10.1016/j.wen.2020.03.008

[3] R. Patel, T. Bhingradiya, A. Deshmukh, V. Gandhi, Response surface methodology for optimization and modeling of photo-degradation of Alizarin cyanine green and acid orange 7 dyes using UV/TiO_2 process, Materials Science Forum 855(2016) 94-104.

[4] S. S. Sable, K. J. Shah, P. C. Chiang, S. L. Lo, Catalytic oxidative degradation of phenol using iron oxide promoted sulfonated-ZrO_2 by advance oxidation processes (AOPs), J. Taiwan Inst. Chem. Eng. 91(2018) 434-440.

[5] Y. Yang, X. Zhou , N. Wei, Research summary on treatment of DNBP-containing industrial wastewater, Modern Chem. Ind., 38(2018) 48-51.

[6] J. Wang, L. Chu, L. Wojnarovits, and E. Takacs, Occurrence and fate of antibiotics, antibiotic resistant genes (ARGs) and antibiotic resistant bacteria (ARB) in municipal wastewater treatment plant: An overview. Sci. Total Environ. 744 (2020) 140997. https://doi.org/10.1016/j.scitotenv.2020.140997

[7] M. Jain, A. Majumder, P.S. Ghosal, and A.K. Gupta, A review on treatment of petroleum refinery and petrochemical plant wastewater: A special emphasis on constructed wetlands. J. Environ. Manage. 272 (2020) 111057. https://doi.org/10.1016/j.jenvman.2020.111057

[8] K. Haider, M.R. Haider, K. Neha, and M.S. Yar, Free radical scavengers: An overview on heterocyclic advances and medicinal prospects. Eur. J. Med. Chem. 204 (2020) 112607. https://doi.org/10.1016/j.ejmech.2020.112607

[9] L. Niu, T. Wei, Q. Li, G. Zhang, G. Xian, Z. Long, and Z. Ren, Ce-based catalysts used in advanced oxidation processes for organic wastewater treatment: A review. J. Environ. Sci.-China 96 (2020) 109-116. https://doi.org/10.1016/j.jes.2020.04.033

[10] K. Avasthi, A. Bohre, M. Grilc, B. Likozar, and B. Saha, Advances in catalytic production processes of biomass-derived vinyl monomers. Catal. Sci. Technol. 10 (2020) 5411-5437. https://doi.org/10.1039/D0CY00598C

[11] A.H.B. Mostaghimi, T.A. Al-Attas, M.G. Kibria, and S. Siahrostami, A review on electrocatalytic oxidation of methane to oxygenates. J. Mater. Chem. A 8 (2020) 15575-15590. https://doi.org/10.1039/D0TA03758C

[12] D.K. Farmer, and M. Riches, Measuring Biosphere-Atmosphere Exchange of Short-Lived Climate Forcers and Their Precursors. Accounts Chem. Res. 53 (2020) 1427-1435. https://doi.org/10.1021/acs.accounts.0c00203

[13] D. Huang, G. Zhang, J. Yi, M. Cheng, C. Lai, P. Xu, C. Zhang, Y. Liu, C. Zhou, W. Xue, R. Wang, Z. Li, and S. Chen, Progress and challenges of metal-organic frameworks-based materials for SR-AOPs applications in water treatment. Chemosphere 263 (2020) 127672. https://doi.org/10.1016/j.chemosphere.2020.127672

[14] P.V. Nidheesh, J. Scaria, D.S. Babu, and M.S. Kumar, An overview on combined electrocoagulation-degradation processes for the effective treatment of water and wastewater. Chemosphere 263 (2020) 127907. https://doi.org/10.1016/j.chemosphere.2020.127907

[15] J. Tan, Z. Li, J. Li, J. Wu, X. Yao, and T. Zhang, Graphitic carbon nitride-based materials in activating persulfate for aqueous organic pollutants degradation: A review

Materials Research Forum LLC
https://doi.org/10.21741/9781644901397-1

on materials design and mechanisms. Chemosphere 262 (2020) 127675. https://doi.org/10.1016/j.chemosphere.2020.127675

[16] U. Mazur, and K.W. Hipps, Single molecule level studies of reversible ligand binding to metal porphyrins at the solution/solid interface. J. Porphyr. Phthalocya. 24 (2020) 993-1002. https://doi.org/10.1142/S1088424620300049

[17] N.Y. Donkadokula, A.K. Kola, I. Naz, and D. Saroj, A review on advanced physico chemical and biological textile dye wastewater treatment techniques. Rev. Environ. Sci. Bio 19 (2020) 543-560. https://doi.org/10.1007/s11157-020-09543-z

[18] J. Li, J. Jiang, S. Pang, Y. Yang, S. Sun, L. Wang, and P. Wang, Transformation of X-ray contrast media by conventional and advanced oxidation processes during water treatment: Efficiency, oxidation intermediates, and formation of iodinated byproducts. Water Res. 185 (2020) 116234. https://doi.org/10.1016/j.watres.2020.116234

[19] Y. Zhu, W. Fan, T. Zhou, and X. Li, Removal of chelated heavy metals from aqueous solution: A review of current methods and mechanisms. Sci. Total Environ. 678 (2019) 253-266. https://doi.org/10.1016/j.scitotenv.2019.04.416

[20] M. Zhang, H. Dong, L. Zhao, D. Wang, and D. Meng, A review on Fenton process for organic wastewater treatment based on optimization perspective. Sci. Total Environ. 670 (2019) 110-121. https://doi.org/10.1016/j.scitotenv.2019.03.180

[21] V.K. Sharma, and M. Feng, Water depollution using metal-organic frameworks-catalyzed advanced oxidation processes: A review. J. Hazard. Mater. 372 (2019) 3-16. https://doi.org/10.1016/j.jhazmat.2017.09.043

[22] A.R. Lado Ribeiro, N.F.F. Moreira, G.L. Puma, and A.M.T. Silva, Impact of water matrix on the removal of micropollutants by advanced oxidation technologies. Chem. Eng. J. 363 (2019) 155-173. https://doi.org/10.1016/j.cej.2019.01.080

[23] M.C. Collivignarelli, A. Abba, M.C. Miino, and S. Damiani, Treatments for color removal from wastewater: State of the art. J. Environ. Manage. 236 (2019) 727-745. https://doi.org/10.1016/j.jenvman.2018.11.094

[24] Z. Tang, Y. Liu, M. He, and W. Bu, Chemodynamic Therapy: Tumour Microenvironment-Mediated Fenton and Fenton-like Reactions. Angew. Chem. Int. Edit. 58 (2019) 946-956. https://doi.org/10.1002/anie.201805664

[25] J. Li, G. Wang, UV-Fenton Oxidation Study on COD Degradation Rate of Wastewater from the Wine Industry in the East of Helan Mountain. Environ. Development, 2019. 10(2019) 107-108.

[26] C. Tang, Q. Chen, F. Lü, J. Lu, S. Du, X. Liu, Three-dimensional electrode electric Fenton oxidation treatment of dye wastewater. Environ. Prot. Chem. Ind. 2019. 1(2019) 28-33.

[27] Y. Zhang, and M. Zhou, A critical review of the application of chelating agents to enable Fenton and Fenton-like reactions at high pH values. J. Hazard. Mater. 362 (2019) 436-450. https://doi.org/10.1016/j.jhazmat.2018.09.035

[28] S.O. Ganiyu, M. Zhou, and C.A. Martinez-Huitle, Heterogeneous electro-Fenton and photoelectro-Fenton processes: A critical review of fundamental principles and application for water/wastewater treatment. Appl. Catal. B-Environ. 235 (2018) 103-129. https://doi.org/10.1016/j.apcatb.2018.04.044

[29] S. Karimifard, and M.R.A. Moghaddam, Application of response surface methodology in physicochemical removal of dyes from wastewater: A critical review. Sci. Total Environ. 640 (2018) 772-797. https://doi.org/10.1016/j.scitotenv.2018.05.355

[30] J. Zhou, J. Zhao, J. Zhang, T. Zhang, M. Ye, and Z. Liu, Regeneration of catalysts deactivated by coke deposition: A review. Chinese J. Catal. 41 (2020) 1048-1061. https://doi.org/10.1016/S1872-2067(20)63552-5

[31] A.C. Mecha, and M.N. Chollom, Photocatalytic ozonation of wastewater: a review. Environ. Chem. Lett. 18 (2020) 1491-1507. https://doi.org/10.1007/s10311-020-01020-x

[32] G. Yu, Y. Wang, H. Cao, H. Zhao, and Y. Xie, Reactive Oxygen Species and Catalytic Active Sites in Heterogeneous Catalytic Ozonation for Water Purification. Environ. Sci. Technol. 54 (2020) 5931-5946. https://doi.org/10.1021/acs.est.0c00575

[33] J. Xiao, Y. Xie, J. Rabeah, A. Brueckner, and H. Cao, Visible-Light Photocatalytic Ozonation Using Graphitic C3N4 Catalysts: A Hydroxyl Radical Manufacturer for Wastewater Treatment. Accounts Chem. Res. 53 (2020) 1024-1033. https://doi.org/10.1021/acs.accounts.9b00624

[34] J. Wang, and H. Chen, Catalytic ozonation for water and wastewater treatment: Recent advances and perspective. Sci. Total Environ. 704 (2020) 135249. https://doi.org/10.1016/j.scitotenv.2019.135249

[35] W. Wang, H. Hu, X. Liu, H. Shi, T. Zhou, C. Wang, Z. Huo, and Q. Wu, Combination of catalytic ozonation by regenerated granular activated carbon (rGAC) and biological activated carbon in the advanced treatment of textile wastewater for reclamation. Chemosphere 231 (2019) 369-377, https://doi.org/10.1016/j.chemosphere.2019.05.175

[36] M. Liu, X. Qi, K. Zhang, M. Wei, C. Jin, Heterogeneous Catalytic Ozone Oxidation Method for Advanced Treatment of Dye Wastewater, Environ. Pollut. Control, 2018. 5(2018) 572-576.

[37] Q. Hu, Y. Hu, D. Gao, Research on the enhanced treatment of wastewater from cellulose ethanol production by ozone oxidation process, Ind. Water Treat. 2(2017) 79-83.

[38] Meng, G., et al., Continuous treatment of biochemical effluent from coking wastewater by ozone oxidation. Environ. Protect. Chem. Ind. 3(2017) 315-319.

[39] D. Li, Z. Xiao, T. Bin Aftab, and S. Xu, Flue Gas Denitration by Wet Oxidation Absorption Methods: Current Status and Development. Environ. Eng. Sci. 35 (2018) 1151-1164. https://doi.org/10.1089/ees.2017.0516

[40] M. Sillanpaa, M.C. Ncibi, and A. Matilainen, Advanced oxidation processes for the removal of natural organic matter from drinking water sources: A comprehensive review. J. Environ. Manage. 208 (2018) 56-76. https://doi.org/10.1016/j.jenvman.2017.12.009

[41] L.F. Garcia, O. L. Lozano, E.V. Lorenzo, M.B. Venta, Y.R. Rodriguez, C.H. Castro, C. Gutierrez Trujillo, I. Fernandez Torres, Review of cytostatic wastewater degradation by ozone and advanced oxidation processes: results from Cuban studies. Environ Rev. 28 (2020) 21-31.

[42] L. Bai, J. Yan, Z. Zeng, Y. Ma, Cavitation in thin liquid layer: A review. Ultra Son. 66 (2020)1083-1092. https://doi.org/10.1016/j.ultsonch.2020.105092

[43] M.J.W. Povey, Applications of ultrasonics in food science - novel control of fat crystallization and structuring. Curr. Opin. Colloid In. 28 (2017) 1-6. https://doi.org/10.1016/j.cocis.2016.12.001

[44] P.N. Amaniampong, and F. Jerome, Catalysis under ultrasonic irradiation: a sound synergy. Curr. Opin. Gre Sus. Chem. 22(2020)7-12. https://doi.org/10.1016/j.cogsc.2019.11.002

[45] F.J. Trujillo, A strict formulation of a nonlinear Helmholtz equation for the propagation of sound in bubbly liquids. Part II: App ultra cavit. Ultra Son. 65 (2020) 1050-1056. https://doi.org/10.1016/j.ultsonch.2020.105056

[46] T. Xu, Z. Cui, D. Li, F. Cao, J. Xu, Y. Zong, S. Wang, A. Bouakaz, M. Wan, S. Zhang, Cavitation characteristics of flowing low and high boiling-point perfluorocarbon phase-shift nanodroplets during focused ultrasound exposures. Ultra Sono. 65 (2020) 1050-1060. https://doi.org/10.1016/j.ultsonch.2020.105060

[47] C.E.H. Tonry, G. Djambazov, A. Dybalska, W.D. Griffiths, C. Beckwith, V. Bojarevics, K.A. Pericleous, Acoustic resonance for contactless ultrasonic cavitation in alloy melts. Ultra Son. 63 (2020)1052-1063. https://doi.org/10.1016/j.ultsonch.2020.104959

[48] X. Xiong, B. Wang, W. Zhu, K. Tian, H. Zhang, A review on ultrasonic catalytic microbubbles ozonation processes: properties, hydroxyl radicals generation pathway and potential in application, Cata. 9 (2019)1014-1027. https://doi.org/10.3390/catal9010010

[49] J. Wang, Z. Wang, C.L.Z. Vieira, J.M. Wolfson, G. Pingtian, S. Huang, Review on the treatment of organic pollutants in water by ultrasonic technology. Ultra Son. 55 (2019) 273-278. https://doi.org/10.1016/j.ultsonch.2019.01.017

[50] X. Huang, H. Hu, S. Li, A. Zhang, Nonlinear dynamics of a cavitation bubble pair near a rigid boundary in a standing ultrasonic wave field. Ultra Son. 64 (2020) 969-987. https://doi.org/10.1016/j.ultsonch.2020.104969

[51] K.M.A. Manmi, W.B. Wu, N. Vyas, W.R. Smith, Q.X. Wang, A.D. Walmsley, Numerical investigation of cavitation generated by an ultrasonic dental scaler tip vibrating in a compressible liquid. Ultra Son. 63 (2020)238-241. https://doi.org/10.1016/j.ultsonch.2020.104963

[52] J. Qian, Y. Li, J. Gao, Z. He, S. Yi, The effect of ultrasonic intensity on physicochemical properties of Chinese fir. Ultra Son, 64 (2020) 104-117. https://doi.org/10.1016/j.ultsonch.2020.104985

[53] E.A. Alenyorege, H. Ma, I. Ayim, C. Zhou, Ultrasound decontamination of pesticides and microorganisms in fruits and vegetables: a review. J. Food Saf Food Qual. 69 (2018) 80-91.

[54] Z. Eren, Ultrasound as a basic and auxiliary process for dye remediation: A review, J. Environ. Manage. 104 (2012) 127-141. https://doi.org/10.1016/j.jenvman.2012.03.028

[55] Y. Chen, H. Zheng, V.N.T. Truong, G. Xie, Q. Liu, Selective aggregation by ultrasonic standing waves through gas nuclei on the particle surface. Ultra Son. 63 (2020)1056-1062. https://doi.org/10.1016/j.ultsonch.2019.104924

[56] T. Xu, X. Lu, D. Peng, G. Wang, C. Chen, W. Liu, W. Wu, T.J. Mason, Ultrasonic stimulation of the brain to enhance the release of dopamine - A potential novel treatment for Parkinsons disease. Ultra Son. 63 (2020)1047-1056. https://doi.org/10.1016/j.ultsonch.2019.104955

[57] J.Zhu, Y. Peng, D. Wan,Ultrasonic-heterogeneous catalytic oxidation for phenol wastewater treatment. J. Food Saf Food Qual.2006(06)783-786.

[58] B.Ren, Experimental study on the treatment of beer wastewater by ultrasound-Fenton oxidation. J. Food Saf Food Qual. 35(2014) 102-104.

[59] B.Ren, C. Zhen, H. Liu,Study on the treatment of washing wastewater by ultrasonic-Fenton oxidation. J. Envir Eng. 36(2015) 101-104.

[60] C. Liu, Y. Chen, C. He, R. Yin, J. Liu, T. Qiu, Ultrasound-Enhanced Catalytic Ozonation Oxidation of Ammonia in Aqueous Solution. Int. J. Env. Res Pub He. 16 (2019). https://doi.org/10.3390/ijerph16122139

[61] Z. Zhou, C. Chen, H. Ren, Homogeneous catalytic enhancement of US/FCME synergistic system for degradation of nitrobenzene. J. Envir Eng. 6(2012)3662-3666.

[62] C. Liu, Y. Chen, C. He, R. Yin, J. Liu, T. Qiu, Ultrasound-Enhanced Catalytic Ozonation Oxidation of Ammonia in Aqueous Solution. Int J Env Res Pub He. 16 (2019)3024-3056. https://doi.org/10.3390/ijerph16122139

[63] S. Arefi-Oskoui, A. Khataee, M. Safarpour, V. Vatanpour, Modification of polyethersulfone ultrafiltration membrane using ultrasonic-assisted functionalized MoS2 for treatment of oil refinery wastewater. Sep Purif Technol. 238 (2020).209-215. https://doi.org/10.1016/j.seppur.2019.116495

[64] M.Sushma, A.Kumari, K. Saroha, Performance of various catalysts on treatment of refractory pollutants in industrial wastewater by catalytic wet air oxidation: A review. J. Environ Manage. 228 (2018) 169-188. https://doi.org/10.1016/j.jenvman.2018.09.003

[65] P.Gautam, S. Kumar, S. Lokhandwala, Advanced oxidation processes for treatment of leachate from hazardous waste landfill: A critical review. J. Clean Prod. 237 (2019)1056-1067. https://doi.org/10.1016/j.jclepro.2019.117639

[66] M.M. M'Arimi, C.A. Mecha, A.K. Kiprop, R. Ramkat, Recent trends in applications of advanced oxidation processes (AOPs) in bioenergy production: Review. Renew Sust Energ Rev. 121 (2020)135-147. https://doi.org/10.1016/j.rser.2019.109669

[67] X .Zeng, L. Xia, Z.Zhong, Study on pretreatment of high concentration synthetic pharmaceutical wastewater by wet oxidation process. Renew Sust Energ Rev. 37(2017)78-80.

[68] S. Wang, X. Wan, Bimetal-wet oxidation method for removing inorganic nitrogen in membrane leachate concentrate of landfill. Renew Sust Energ Rev.43(2017) 62-66.

[69] C. Wu, L. Zhang, K.Zhong, Study on Catalytic Wet Oxidation for Simulating Space Station Condensate Wastewater and Sanitary Wastewater. Sep Purif Technol.25(2015) 152-167.

[70] R. Castaldo, A new route for low pressure and temperature CWAO: A PtRu/MoS2 hyper-crosslinked nanocomposite. Nanomaterials, 29(2019)142-163. https://doi.org/10.3390/nano9101477

[71] E.G. Vaschetto, M.I. Sicardi, V.R. Elias, G.O. Ferrero, P.M. Carraro, S.G. Casuscelli, G.A. Eimer, Metal modified silica for catalytic wet air oxidation (CWAO) of glyphosate under atmospheric conditions. Adsorption. 25 (2019) 1299-1306. https://doi.org/10.1007/s10450-019-00090-w

[72] P. Wang, Y.N. Liang, Z. Zhong, X. Hu, Nano-hybrid bimetallic Au-Pd catalysts for ambient condition-catalytic wet air oxidation (AC-CWAO) of organic dyes. Sep Purif Technol. 233 (2020)335-348. https://doi.org/10.1016/j.seppur.2019.115960

[73] A. Al-Atta, J. Sierra-Pallares, T. Huddle, E. Lester, A potential co-current mixing reactor design for supercritical water oxidation. J Supercrit Fluid. 158 (2020)234-256. https://doi.org/10.1016/j.supflu.2019.104708

[74] D. Zhang, A.K. Luther, P. Clauwaert, P. Ciccioli, F. Ronsse, Assessment of carbon recovery from solid organic wastes by supercritical water oxidation for a regenerative life support system. Environ Sci Pollut R. 27 (2020) 8260-8270. https://doi.org/10.1007/s11356-019-07527-3

[75] F. Zhang, J. Chen, C. Su, S. Chen, Z. Chen, Y. Ding, Combined effects of protective film and oxidant on the performance of the supercritical water oxidation system with a film protective reactor: A simulation study. Process Saf Environ. 131 (2019) 268-281. https://doi.org/10.1016/j.psep.2019.09.026

[76] D. Zhang, S. Ghysels, F. Ronsse, Effluent recirculation enables near-complete oxidation of organics during supercritical water oxidation at mild conditions: A proof of principle. Chemosphere. 250 (2020) 126-146. https://doi.org/10.1016/j.chemosphere.2020.126213

[77] S. Zhang, Z. Zhang, R. Zhao, J. Gu, J. Liu, B. Ormeci, J. Zhang, A Review of Challenges and Recent Progress in Supercritical Water Oxidation of Wastewater. Chem Eng Commun. 204 (2017) 265-282. https://doi.org/10.1080/00986445.2016.1262359

[78] Z, Xu, X. Zhang, R. Zhou. Research Status and New Progress of Advanced Wastewater Oxidation Technology. Chem Eng Commun. 44(2018)7-10.

[79] Q. Li, X. Zhou, and J. Zhang, Supercritical water oxidation treatment of simulated dye wastewater. Chem Eng. 44(2018)10-14.

[80] Li, j., et al. Study on the treatment of landfill leachate by the combination of Fenton method and supercritical water oxidation method. Chem Eng J. 41(2015)89-92.

[81] Z. Yan, M. Zhang, Y. Han, J. Zhang, Supercritical water oxidation for p-tert-butylcatechol degradation in wastewater. Chem Eng. 44 (2016) 70-74.

[82] J. Li, S. Wang, Y. Li, L. Wang, T. Xu, Y. Zhang, Z. Jiang, Supercritical water oxidation of semi-coke wastewater: Effects of operating parameters, reaction mechanism and process enhancement. Sci Total Environ. 710 (2020)146-165. https://doi.org/10.1016/j.scitotenv.2019.134396

[83] Z. Cheng, Q. Chen, S. Cervantes, Q. Tang, X. Gao, Y. Tan, S. Liu, Y. Ma, Z. Shen, Two-dimensional and Three-dimensional quantitative structure-activity relationship models for the degradation of organophosphate flame retardants during supercritical Water oxidation.. J Hazard Mater. 394 (2019) 121811. https://doi.org/10.1016/j.jhazmat.2019.121811

[84] D. Xu, S. Wang, C. Huang, X. Tang, Y. Guo, Transpiring wall reactor in supercritical water oxidation. Chem Eng Res Des. 92 (2014) 2626-2639. https://doi.org/10.1016/j.cherd.2014.02.028

[85] D. Xu, C. Huang, S. Wang, G. Lin, Y. Guo, Salt deposition problems in supercritical water oxidation. Chem Eng J. 279 (2015) 1010-1022. https://doi.org/10.1016/j.cej.2015.05.040

[86] D. Zhi, Y. Lin, L. Jiang, Y. Zhou, A. Huang, J. Yang, L. Luo, Remediation of persistent organic pollutants in aqueous systems by electrochemical activation of persulfates: A review. J Environ Manage. 260 (2020)165-175. https://doi.org/10.1016/j.jenvman.2020.110125

[87] C. Liu, J. Li, H. Pang, Metal-organic framework-based materials as an emerging platform for advanced electrochemical sensing. Coordin Chem Rev. 410 (2020)16523. https://doi.org/10.1016/j.ccr.2020.213222

[88] R. Gao, D. Yan, Recent Development of Ni/Fe-Based Micro/Nanostructures toward Photo/Electrochemical Water Oxidation. Adv Energy Mater. 10 (2020)123654. https://doi.org/10.1002/aenm.201900954

[89] E.R. Corson, E.B. Creel, R. Kostecki, B.D. McCloskey, J.J. Urban, Important Considerations in Plasmon-Enhanced Electrochemical Conversion at Voltage-Biased Electrodes.. Iscience. 23 (2020) 100911. https://doi.org/10.1016/j.isci.2020.100911

[90] P.C. Lindholm-Lehto, J.S. Knuutinen, H.S.J. Ahkola, S.H. Herve, Refractory organic pollutants and toxicity in pulp and paper mill wastewaters. Environ Sci Pollut R. 22 (2015) 6473-6499. https://doi.org/10.1007/s11356-015-4163-x

[91] L.L. Huang, J.F. Liu, B. Sun, N. Zhang, Y.Q. Tang, Y.J. Feng, Experimental Study on Papermaking Wastewater by Advanced Electrochemical Oxidation Method, Advanced Materials Research. Trans Tech Publ. 145(2013)1699-1703. https://doi.org/10.4028/www.scientific.net/AMR.726-731.1699

[92] P. Aravind, H. Selvaraj, S. Ferro, G.M. Neelavannan, M. Sundaram, A one-pot approach: Oxychloride radicals enhanced electrochemical oxidation for the treatment of textile dye wastewater trailed by mixed salts recycling. J Clean Prod. 182 (2018) 246-258. https://doi.org/10.1016/j.jclepro.2018.02.064

[93] Y. Yavuz, A.S. Koparal, U.B. Ogutveren, Treatment of petroleum refinery wastewater by electrochemical methods. Desalination. 258 (2010) 201-205. https://doi.org/10.1016/j.desal.2010.03.013

[94] C.R. Wang, Z.F. Hou, M.R. Zhang, J. Qi, J. Wang, Electrochemical Oxidation Using BDD Anodes Combined with Biological Aerated Filter for Biotreated Coking Wastewater Treatment. J Chem-Ny. 56(2015)135-146. https://doi.org/10.1155/2015/201350

[95] F. Agustina, A.Y. Bagastyo, E. Nurhayati, Electro-oxidation of landfill leachate using boron-doped diamond: role of current density, pH and ions. Water Sci Technol. 79 (2019) 921-928. https://doi.org/10.2166/wst.2019.040

[96] L. Lei, D. Huang, C. Zhou, S. Chen, X. Yan, Z. Li, W. Wang, Demystifying the active roles of NiFe-based oxides/(oxy)hydroxides for electrochemical water splitting under alkaline conditions. Coordin Chem Rev. 408 (2020)496-512. https://doi.org/10.1016/j.ccr.2019.213177

[97] M. Pirsaheb, H. Hossaini, H. Janjani, An overview on ultraviolet persulfate based advances oxidation process for removal of antibiotics from aqueous solutions: a systematic review. Desalin Water Treat. 165 (2019) 382-395. https://doi.org/10.5004/dwt.2019.24559

[98] Y. Ahn, E. Yun, Heterogeneous metals and metal-free carbon materials for oxidative degradation through persulfate activation: A review of heterogeneous catalytic activation of persulfate related to oxidation mechanism. Korean J Chem Eng. 36 (2019) 1767-1779. https://doi.org/10.1007/s11814-019-0398-4

[99] Y. Pang, K. Luo, L. Tang, X. Li, J. Yu, J. Guo, Y. Liu, Z. Zhang, R. Yue, L. Li, Carbon-based magnetic nanocomposite as catalyst for persulfate activation: a critical

review. Environ Sci Pollut R. 26 (2019) 32764-32776. https://doi.org/10.1007/s11356-019-06403-4

[100] S. Xiao, M. Cheng, H. Zhong, Z. Liu, Y. Liu, X. Yang, Q. Liang, Iron-mediated activation of persulfate and peroxymonosulfate in both homogeneous and heterogeneous ways: A review. Chem Eng J. 384 (2020)147258. https://doi.org/10.1016/j.cej.2019.123265

[101] H. Li, Z. Han, Y. Qian, X. Kong, P. Wang, In Situ Persulfate Oxidation of 1,2,3-Trichloropropane in Groundwater of North China Plain. Int J Env Res Pub He. 16 (2019)136544. https://doi.org/10.3390/ijerph16152752

[102] H. Kusic, I. Peternel, N. Koprivanac, A.L. Bozic, Iron-Activated Persulfate Oxidation of an Azo Dye in Model Wastewater: Influence of Iron Activator Type on Process Optimization. J. Environ Eng-Asce. 137 (2011) 454-463. https://doi.org/10.1061/(ASCE)EE.1943-7870.0000347

[103] Z. Huang, Z. Ji, Y. Zhao, J. Liu, F. Li, J. Yuan, Treatment of wastewater containing 2-methoxyphenol by persulfate with thermal and alkali synergistic activation: Kinetics and mechanism. Chem Eng J. 380 (2020)3465-3476. https://doi.org/10.1016/j.cej.2019.122411

[104] W. Yang, G. Liu, Y. Chen, D. Miao, Q. Wei, H. Li, L. Ma, K. Zhou, L. Liu, Z. Yu, Persulfate enhanced electrochemical oxidation of highly toxic cyanide-containing organic wastewater using boron-doped diamond anode.. Chemosphere. 252 (2020) 126499. https://doi.org/10.1016/j.chemosphere.2020.126499

[105] Y. Wang, Z. Gu, S. Yang, A. Zhang, Activation of persulfate by microwave radiation combined with FeS for treatment of wastewater from explosives production. Environ Sci-Wat Res. 6 (2020) 581-592. https://doi.org/10.1039/C9EW00803A

[106] Z. Zhou, X. Liu, K. Sun, C. Lin, J. Ma, M. He, W. Ouyang, Persulfate-based advanced oxidation processes (AOPs) for organic-contaminated soil remediation: A review. Chem Eng J. 372 (2019) 836-851. https://doi.org/10.1016/j.cej.2019.04.213

Materials Research Foundations **102** (2021) 48-67 — https://doi.org/10.21741/9781644901397-2

Chapter 2

Electrochemical Oxidation of Perfluorooctanoic Acid (PFOA) from Aqueous Solution using Non-Active Ti/SnO_2-Sb_2O_5/PbO_2 Anodes

Seema Singh[1,2], Kinjal J. Shah[3], Nidhi Mehta[1,4], Vimal Chandra Srivastava[5], Shang-Lien Lo[1,6,*]

[1]Graduate Institute of Environmental Engineering, National Taiwan University, 71 Chou-Shan Rd., Taipei, Taiwan, ROC

[2]Omvati Devi Degree College Bhalaswagaj, Haridwar, Uttarakhand, India

[3]College of Urban Construction, Nanjing Tech University, Fuzhu, Nanjing, China

[4]Department of Biochemistry, University of Washington, Seattle, WA, 98195, USA

[5]Department of Chemical Engineering, Indian Institute of Technology, Roorkee, Roorkee 247667, Uttarakhand, India

[6]Water Innovation, Low Carbon and Environmental Sustainability Research Center, National Taiwan University, Taipei 10617, Taiwan

* sllo@ntu.edu.tw (S.-L. Lo)

Abstract

In this study, electrochemical oxidation of perfluorooctanoic acid (PFOA, $C_7H_{15}CO_2H$) from aqueous solution was examined in terms of PFOA and total organic carbon (TOC) removal by using Ti/SnO_2-Sb_2O_5/PbO_2non-active electrodes. The effects of operating parameters: initial pH (pH_o), current density (j), and electrolyte concentration (m) at different time intervals were examined. Specific energy consumption (SEC) was used to determine the process proficiency. The C-C bond between C_7F_{15} was first cleaved and thendegraded into fluoride ions (F^-) and short carbon-chain per-fluorinated carboxylic acids (PFCAs) ((~C2−C7) such as perfluoroethanoic acid (PFEA: $C_2F_5CO_2H$), perfluoropropanoic acid (PFPA: $C_3F_7CO_2H$), perfluorobutanoic acid (PFBA: $C_4F_9CO_2H$), perfluoropentanoic acid (PFPeA: $C_5F_{11}CO_2H$), perfluorohexanoic acid (PFHxA: $C_6F_{13}CO_2H$), perfluoheptanoic acid (PFHpA: $C_7F_{14}CO_2H$). These intermediates by-products were determined using the gas chromatograph-mass spectrometry (GC/MS) analysis. The rate of PFOA decomposition was followed the pseudo-first-order kinetics. About 82%TOC and 91% PFOA removals were formed at the optimal condition of pH_o = 3.58, j=168.34 Am^{-2}, and m = 250 mgL^{-1} at 120 min of electrolysis with SEC = 593 kWh/kg

TOC. A plausible degradation mechanism was also proposed at the optimal treatment condition.

Keywords

Perfluorooctanoic Acid, Electrochemical Oxidation, Specific Energy Consumption, Reaction Kinetics, Degradation Mechanism

Contents

1. Introduction

Polyfluoroalkyl substances (PFASs) and perfluorooctanoic acid (PFOA) are a group of chemicals that possess a complete fluorinated alkyl chain of various lengths which consists ofan acidic head group i.e. carboxylic acid, phosphoric acid and sulfonic acid. These are extremely stable chemicals withhigher C-F bond energy ((531.5 kJ mol^{-1}), which givesthem higher thermal resistance and surface activity [1]. Along with PFASs, PFOA is broadly used various industrial products such as sports clothing, lubricants, extreme weather military uniforms, firefighting foams, waterproof breathable fabrics, motor oil additives and medical activities as metal coatings, surfactants, fire retardants, and surface treatments [2,3]. Owing to their synthesis and wide use of applications, PFOA has been regularly identified in the natural environment, air, water, wildlife, human serum, and solid waste [4,5]. After discharge, it cansimplybe transported into the water system. Environmental monitoring has exposed their persistence, bioaccumulation, and acute toxicity to aquatic organisms [5-8].

As a result, it is necessary to develop a systematic process to eliminate the PFOA from the wastewater. Some treatment methods including adsorption, coagulation, UV irradiation, ultrasonic irradiation, electrochemical oxidation, and zero-valent iron (ZVI) reduction were employed [5,7-12]. Although some of these processes need acute conditions of treatment because of comparatively unsuccessful PFOA removal and limit to their large-scale application [11-15]. Therefore, it is a great challenge to develop an environmentally friendly and cost-effective technology such as high mineralization and complete degradation of PFOA.

In contract of the above processes, the electrochemical oxidation of pollutants removal from aqueous medium has become a promising technologybecause of its higher oxygen evolution potential (OEP) towards anode materials, excellent oxidation ability, electron transfer capacity, mild treatment conditions, hydroxyl radical ($^{\bullet}OH$) formation, and environmental compatibility. Recently, some studies have been performed on PFOA degradation with electrochemical oxidation [1,16,17] and it could be effectively degraded with "non-active" anodes, such as niobium (Nb), tungsten (W), and tantalum (Ta) doped boron-doped diamond (BDD), lead dioxide (PbO_2) electrodes, and tin oxide (SnO_2) even at higher concentrations [18] than "active anodes" including Ti/RuO_2, Ti/IrO_2, Ti/ IrO-RuO_2, Pt and graphite electrodes due to higher OEP, higher electron transfer ability, and higher $^{\bullet}OH$ generation rate [1,19,20]. Previous reports have examined the EO performance for PFOA with non-active anodes and generally followed in the following order: BDD > SnO_2–Sb > PbO_2-Ce [21].

Although BDD has superior performance due to its long life, higher OEP, and excellent chemical and electrochemical stability but higher cost and particularly hard to find the suitable substrate for the deposition limit large –scale uses of BDD electrode. Nb, Ta, and W substrates are too costly; silicon (Si) is extremely fragile and have lesser conductivity; Ta (Ti/BDD) is a choice as a substrate, however service life is comparatively small [1,16,17]. Thus, in search of a low-cost industrialized electrode for the PFOA degradation shows a very significant. Potential electrodes, particularity Sb doped Ti/SnO_2-Sb/PbO_2 anodes, have various benefits such as high OEP (~1.9 V versus SHE), long life time, easy preparation, cost efficiency, and high chemical stability [1,19,20] confirmed the 73.9% defluorination and 98.8% degradation of PFOA were formed over Ti/SnO_2–Sb anode at the optimum treatment conditions of j= 40 $mAcm^{-2}$,pH=5, electrolyte concentration of 10 mM NaClO4 and keeping the distance between the electrode plate about 1.0 cm for 100 mg L^{-1} PFOA initial concentration after 90 min electrolysis. But, Zhao et al. [20] have confirmed that 20% PFOA removal efficiency decrease at Ti/SnO_2-Sb anode in presence of Na_2SO_4 electrolyte because SO_4^{2-} ions adsorbed at the surface of anode and its active sites was occupied which results decrease in the concentration of $^{\bullet}OH$ generation.

This study aimed to examine the electrochemical oxidation (EO) of PFOA using Ti/SnO_2-Sb_2O_5/PbO_2 anodes in aqueous solution. We carried out the experiments to determine the effect of various influencing factors such as solution pH (pH_o), current density (j), and sodium chloride (NaCl) as an electrolyte concentration (m) for PFOA and TOC removal. Specific energy consumption (SEC) during each parameter optimization was used to detect the process performance of EO. To evaluate the PFOA degradation rate constant, kinetic studies of PFOA removal was determine. In addition, intermediate by-products such as organic and inorganic byproducts were identified during PFOA electrolysis, and finally, a plausible degradation mechanism of PFOA removal during EO was suggested with the help of intermediates byproducts identification.

2. Materials and methods

2.1 Materials

All chemicals of analytical grade were used in this study. Hydrochloric acid (HCl), sodium bi-carbonate ($NaHCO_3$), sodium carbonate (Na_2CO_3), and sodium hydroxide (NaOH) were purchased from Nacalai Tesque. The synthetic solution of PFOA (Sigma Aldrich Chemical, USA) was prepared by dissolving its suitable amount of deionized/ultrapure water. NaOH and HCl were used to adjust the solution's pH.

2.2 Electrode preparation and characterization

Rectangular shape of Ti plates with a dimension of 2.5 cm × 2.0 cm and 1 mm thickness were prepared by immersed it into a freshly prepared 40 wt% NaOH solution at 333K for 2h to remove the grease, and then plates were etched for 2h by boiling with 10wt% oxalic acid to generate a grey black surface with regular roughness. Later, Ti/SnO_2–Sb_2O_5 interlayer was prepared by thermal decomposition method. In brief, the precursor solution was prepared by mixing the mole ratio 9:1 of $SnCl_4.5H_2O$ and $SbCl_3$ in butanol and hydrochloric acid. After that, paint solution was applied on pretreated Ti plates with a brush, which was dried at 393K for 10min in the oven, and then sintered in the muffle furnace for the coating at 773K for 10 min. This process was revised ten times to generate the Ti/SnO_2–Sb electrodes, and the final baking was kept at 773K for 1h. At the end, various interlayers were attained. Afterwards, β-PbO_2 films were electrodeposited on the as-prepared interlayers of Ti/SnO_2–Sb. For this purpose the β-PbO_2 was produced from 0.1M HNO_3 acidic medium containing 0.5M $Pb(NO_3)_2$ and 40 mM NaF. The Ti plate and interlayer were employed as an anode and cathode, respectively. Both anode and cathode were perpendicularly put in the electrodeposition solution with a fixed distance of 1.0 cm. The β-PbO_2 films were electrodeposited after 2h with a constant j of 200 A cm^{-2} at 333K. At last, the different interlayers of β-PbO_2 electrodes were formed.

2.3 Electrochemical experiments

The EO experiments of PFOA were conducted in a double layered cylindrical batch reactor having 1.5 L capacity of wastewater treatment. Each experiment was conducted with a constant direct current (DC) power supply (GPR-30H10D) connected with two Ti/SnO_2-Sb_2O_3/β-PbO_2 electrodes and the distance between each electrodes was maintained about 1 cm. The base of electrodes was kept 5 cm above from EC reactor bottom for the trouble-free stirring of solution at 400–500 rpm. The batch tests were carried out at room temperature and the voltage varied between the ranges of 1–15 V. All experiments were conducted with 1 L PFOA solution in a batch reactor. The conductivity of solution was maintained by using the NaCl electrolyte and the dilute solution of NaOH and HCl were used to adjust the solution pH. The EO experiments were conducted to optimize the various operating parameters within the various ranges such as initial pH (pH_o) of solution 3.8–10, current density (j): 56.31–225.25 Am^{-2}, electrolyte concentration (m): 150–300 mgL^{-1} and treatment time of 180 min. The treated solution was collected at different time intervals of each experiment and centrifuged then filtered prior to know the residual TOC and PFOA concentration.

2.4 Instrumental analysis

The initial and final concentration of PFOA was determined with a UV- spectrophotometer (UV-1800, Shimadzu) and TOC of the solution was measured in a TOC analyzer (1030 TOC analyzer, USA) equipped with OI Analytical model. Before TOC detection, all samples were acidified, filtered with 0.22-μm syringe filter and then nitrogen gas spurge in the solution to remove the inorganic carbon before the analysis. Gas chromatograph-mass spectra (GC-MS) analysis was used to detect the intermediate by-products. For GC-MS analysis, GC and mass spectrometer (MS, Saturn 2100T) equipped with a capillary column and electron spray source respectively to extract the intermediate by-products of PFOA by using the 1:1 ratio of 10mL of di-chloro-methane (DCM) mixed with 10 mL sample of treated PFOA solution. This procedure was repeated thrice, and then anhydrous sodium sulfate (dehydrating agent) was used to dehydrate the sample. The complete range of mass spectra was obtained between the 100 to 450 m/z by using the following program.

The desolvation and source temperature were keptat 550°C and 120°C.The capillary potential and cone voltage were 4.5 kV and 35 V, respectively. N_2 gas pressure was 0.8 MPa at 1.0 s scan time and an average of 90 scans were used to obtain the mass spectra. NIST library was used to determine the intermediate by-products. Fluoride ions concentration (F^-) was measured by using an ion chromatograph system (ICS-3000, Dionex, USA) with a separation column (150 × 4.0 mm) couple with conductivity detector. A mixture of $NaHCO_3$ (1.0 mM) and Na_2CO_3 (3.2 mM) was used as a mobile phase at 2 mL min^{-1} flow rate for anion measurement, and 50 μL was kept the sample injection volume.

The TOC and PFOA removal efficiencies were determined by using the following relationships (Eqs. 1-4):

$$\text{TOC removal (\%)} = \frac{(TOC_o - TOC_f)}{TOC_o} \times 100 \quad (1)$$

$$\text{PFOA removal (\%)} = \frac{PFOA_o - PFOA_f}{PFOA_o} \times 100 \quad (2)$$

where TOC_o/TOC_f and $PFOA_o/PFOA_f$ are the TOC and PFOA concentrations (mgL^{-1}) at t=0 and after t=t.

The amount of energy absorbed (in kWh) per kg TOC removed in terms of specific energy consumption (SEC) was measured by using the following expression:

$$\text{SEC} = \frac{V_{ap} I \Delta t}{(TOC_i - TOC_f) V_r} \times 100 \quad (3)$$

where TOC_i/TOC_f are the initial and final TOC (mg L^{-1}), V_{ap}, I and t are the applied voltage (V), applied current (A) and electrolysis time (h), respectively.

During electro-oxidation, the removal efficiency is directly proportional to the oxidants produced during electrolysis and initial PFOA concentration. Therefore, the kinetics of PFOA and TOC removal can describe via the pseudo first-order kinetic mode

$$\frac{d}{dt}[C] = k\,[C] \quad (4)$$

The value of k rate constant (min $^{-1}$) for TOC reduction (mg L^{-1}) was determined by plotting the graph between ln (C_o/C_t) versus treatment time t (min).

3. Results and discussion

3.1 Effect of pH

Solution pH greatly affected the PFOA and TOC removal efficiencies because it influences the concentration of OH production, life span of anodes, and OER which consequently affects the EO efficiency of PFCs removal[1,22]. Fig.1a and bshows the pH effect from 3.8-9 at j=166.67 Am^{-2}, initial PFOA concentration 100 mg L^{-1} and NaCl concentration of 250 mg L^{-1}. PFOA and TOC removal efficiencies were found to be higher in the acidic range of solution pH because higher OER and OH radicals' production rate increased [19,23].In acidic medium, the low range the solution pH restricts the O_2 evolution reaction which increases the degradation of PFOA to some degree. Fig. 1a and b shows the maximum PFOA and TOC reduction at pH=3.58, but the removal efficiencies were decreased from the pH of 5 to 9. This is because the active sites of an electrodes adsorbed more OH radicals in the alkaline medium, these radicals progressively frighten with $CF_3(CF_2)_6CO_2^-$ anion via the decarboxylation reaction, whereas the acidic pH range helps in PFOA removal.[19] These results confirm that the acidic pH promote the PFOA degradation with Ti/SnO_2-Sb electrode than alkaline pH. The PFOA degradation rate was found 2 times higher at initial pH of 5 than that of pH=11. Similarly, Zhao et al. [23] was observed that the PFOA degradation rate decreased with Si/BDD anode with increasing the pH conditions in the following sequence of pH 3> pH 9>pH 12. Fig. 1c shows the pH effect on SEC. The SEC values were found maximum for pH=3.8 after 180 min of electrolysis, while the value of SEC increased from 660-743.24 kWh/kg TOC with increasing the solution pH from 3.8 to 9. Fig. 1d shows the values of k and R^2 decrease from 14×10^{-3} to 9×10^{-3} when solution pH was increased from pH =3.8 to pH=9, signifying that acidic pH more favorable to increasing the PFOA degradation and higher OH radicals generation.

3.2 Effect of applied current density

The j value isan important operating parameter that influences the EO of PFOA in aqueous solution; it regulates the OH radical generation and electron transfer on the surface of electrode [19,22,23]. Fig. 2a and b illustrates the effect of j on PFOA degradation and TOC reduction in the range of 56.31-225.25 Am^{-2}.

Materials Research Foundations **102** (2021) 48-67 https://doi.org/10.21741/9781644901397-2

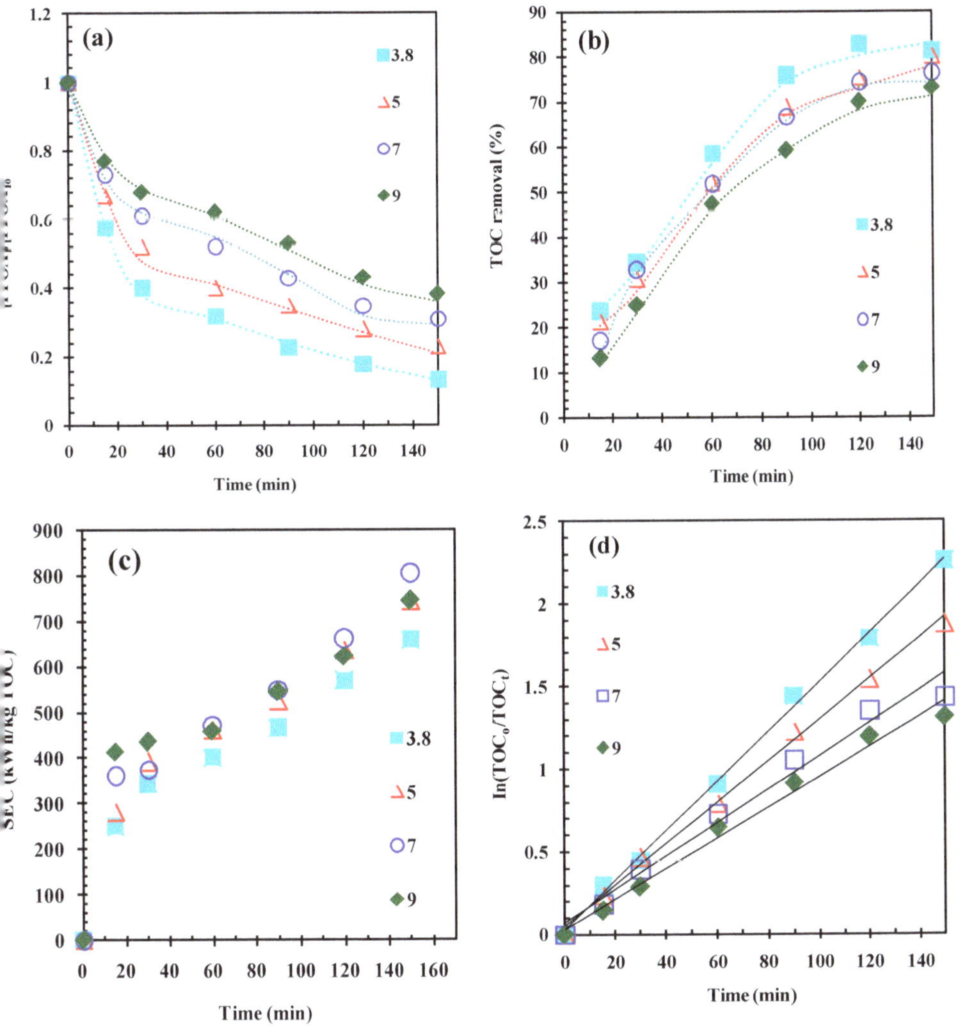

Figure 1: Effect of j on PFOA degradation: (a) PFOA removal, (b) TOC removal efficiency, (c) kinetics studies, and (d) SEC (kWh/kg TOC) at pH 3.8, m = 250 mgL^{-1} and C = 100 mgL^{-1} at different time intervals.

Materials Research Forum LLC
https://doi.org/10.21741/9781644901397-2

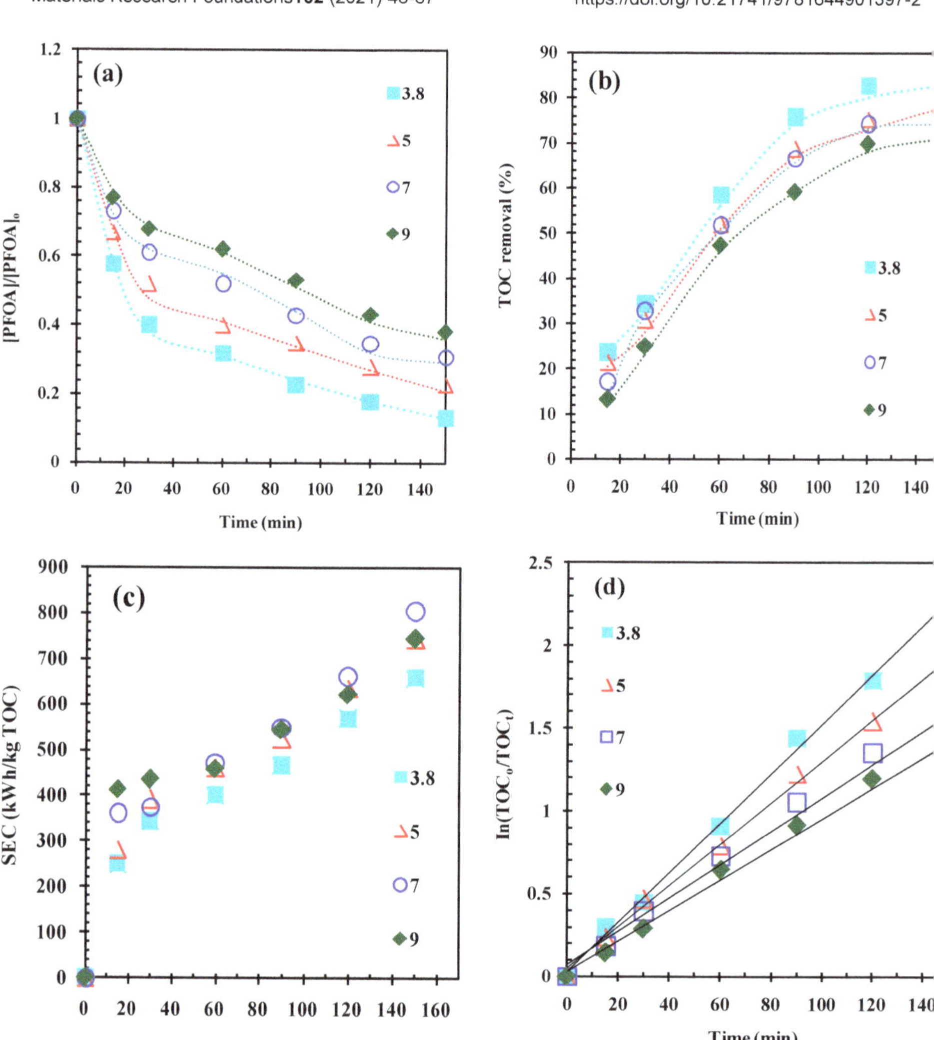

Fig 2: *Effect of solution pH on PFOA degradation: (a) PFOA removal, (b) TOC removal efficiency, (c) kinetics studies, and (d) SEC (kWh/kg TOC) at j=168.9 Am*$^{-2}$*, m = 250 mgL*$^{-1}$ *and C = 100 mgL*$^{-1}$ *at different time intervals.*

Table 1:TOC removal and kinetic characteristics of PFOA degradation after 150 min of electrolysis for different parameters.

Parameters	**TOC removal (%)**	**Kinetic Study**	
		K (min^{-1})	R^2
j ($A\ m^{-2}$)	Other conditions: pH = 3.5, C_o = 100 mg L^{-1}, m = 1.5 g L^{-1}		
56.31	62	7×10^{-3}	0.963
112.62	75	12×10^{-3}	0.995
168.94	82	14×10^{-3}	0.996
225.25	87	15×10^{-3}	0.997
Solution pH_o	Other conditions: j = 168.94 A m^{-2}, C_o = 100 mg L^{-1}, m = 1.5 g L^{-1}		
3.8	82	14×10^{-3}	0.963
5	75	12×10^{-3}	0.996
7	74	10×10^{-3}	0.997
9	68	9×10^{-3}	0.986
m (mgL^{-1})	Other conditions: j = 168.94 A m^{-2}, C_o = 100 mg L^{-1}, pH_o = 3.8		
150	67	9×10^{-3}	0.994
200	73	10×10^{-3}	0.995
250	82	12×10^{-3}	0.991
350	89	14×10^{-3}	0.996

The low j=56.31Am^{-2} could not sufficient to stimulate the PFOA degradation because low potential compared to oxidation potential of PFOA degradation and also a poor OH radicals concentration were produced by the anodes [22]. For example, 6.2% PFOA decomposition was attained at low j= 1.42 mA cm^{-2} with Ti/ SnO_2–Sb_2O_5–Bi_2O_3 electrode [22]. However, PFOA and TOC removal efficiencies were increased with increasing the j from 56.31-225.25 Am^{-2}. For j=168.94 Am^{-2}, the PFOA and TOC removal were maximum about more than 90% and 82%, respectively (Fig. 2a and b) because rate of OH radical formation rate increased with increasing the j with t (Eqns. 5 and 6) that effectively participates in the oxidation reaction and PFOA mineralization (Eqn. 7). Similarly, H_2O_2 production rate also increased via the cathodic reaction of oxygen reduction (Eqns. 8 and 9) at higher j, which in turn raises the electron generation rate and helps in the PFOA degradation [19,23]. The PFOA decomposition rate considerably increased from 76.9% to 98.8% with Ti/SnO_2–Sb anode when the j values were increased from 5 to 40 mA cm^{-2} after 90 min of electrolysis. Fig. 2a and b shows a slightly changed in PFOA oxidation at higher j of 225.25 Am^{-2} because O_2 discharge reaction and diffusion control suppressed the PFOA degradation.

$$H_2O \rightarrow OH^{\bullet} + H^+ + e \quad E^o(OH/H_2O)\ 2.80\ V\ \text{versus SHE} \tag{5}$$

$$2H_2O \rightarrow O_2 + 4H^+ + 4e \quad (6)$$

$$OH^{\bullet} \rightarrow 1/2O_2 + H^+ + e \quad (7)$$

$$O_{2(g)} + 2H^+ + 2e \rightarrow H_2O_2 E^o(O_2/H_2O_2)\ 1.23\ V \quad (8)$$

$$4H_2O + 2O_{2(aq)} + 4e \rightarrow H_2O_2 + 4OH^- \quad (9)$$

Moreover, higher j leads to higher energy consumption and shorter electrode life. Thus, it is very essential to select an appropriate j for PFOA removal in practical application. Fig. 2c has shown the difference of SEC at different j and various time intervals. An increase in j and t, the SEC values increased because the PFOA conversion into stable intermediates by-products increased that oppose the more oxidation behind a definite time of treatment, therefore the SEC value was increased. The SEC value for higher PFOA degradation and TOC reduction at j=168.94 Am^{-2} was found to be 660.23 kWh/ kg TOC^{-1}. Fig. 2d shown straight lines of PFOA degradation with time at different j because the PFOA degradation kinetics were followed the pseudo-first order at different j values. Table 1 illustrates the different values of k and R^2 for TOC reduction at different j. The k and R^2 values were increased from 7×10^{-3} to 15×10^{-3} and 0.963-0.997, respectively at j from 56.31-225.25 Am^{-2}.

3.3 Effect of electrolyte concentration

An electroconductive medium of electrolysis was provided by using the supporting electrolytes to minimize the resistance of electrochemical reactor and the voltage drop [19,23]. A better pollutant removal was occurred with higher conductivity which leads to the quick electrons transport [1]. Fig. 3a and bshows the PFOA and TOC removal efficiencies, these values were increased with increasing the amount of NaCl electrolyte concentration. The concentration of NaCl was increased from 150 to 250 mgL^{-1} then the PFOA and TOC removal were also increased from 70 to 95% and 67 to 84%, respectively, however, at a certain level of NaCl concentration which is about 300 mgL^{-1} no further removal were formed. Fig.3c shows the value of SEC increased with different electrolyte concentration. However, a certain change was formed with time for a particular time interval. This is the reason of maximum TOC reduction was attained between 30-150 min of electrolysis, therefore the PFOA degradation was significantly influenced by the dose of electrolyte. Higher the dose of NaCl facilitates electron transfer between the pollutant

Materials Research Forum LLC
https://doi.org/10.21741/9781644901397-2

and the electrode via offering the active species such as OH radicals and chlorine. Chloride ions (Cl^-) were produced in via the electrochlorination when NaCl used as an electrolyte.

The main chlorination reactions formed during EO via the anodic and cathodic reactions of Cl^- ions with electro-generated $^{\bullet}OH$ radicals are discuss (Eq. 11-14) [24]. At anode, the Cl^-ions oxidized directly into soluble Cl_2 (Eq.10):

$$Cl^- + 2e \rightarrow 2Cl_{2(aq)}\ (E^o = 1.36\ V\ versus\ SHE) \quad (10)$$

Even though the Cl_2 and H^+ concentrations are not sufficient, the Cl_2 potentially contributes in the formation of other active chloro species such as ClO^- (Eq.11), HOCl (Eq.12), ClO_2 (Eq.13) and $Cl^{\bullet}$ (Eq.14) by the disproportionate reactions which could effective enhanced the PFOA degradation.

$$Cl_2 + 2HO^- \rightarrow H_2O + ClO^- + Cl^- \quad (11)$$

$$Cl_{2(aq)} + H_2O \rightarrow HOCl + Cl + H^+\ (E^o = 0.47\ V\ versus\ SHE) \quad (12)$$

$$Cl_2 + 4H_2O \rightarrow 2ClO_2 + 8H^+ + 8\ e^- \quad (13)$$

$$Cl_2 + 2\ e^- \rightarrow 2Cl^- \quad (14)$$

Moreover, when electrolyte concentration increased, the conductivity of solution also enhanced because increasing the electron transfer rate which enhanced the PFOA degradation rate [22]. The straight lines of PFOA degradation with time at different m in Fig.3d depicted the degradation was followed the pseudo-first order kinetics. Table 1 shows the values of k and R^2 for TOC reduction at various m. The k and R^2 values were increased from 9×10^{-3} to 14×10^{-3} and 0.994-0.996, respectively at m from 150 to 300 mg $L^{-1}$and illustrated the about 67 to 89% TOC reduction.

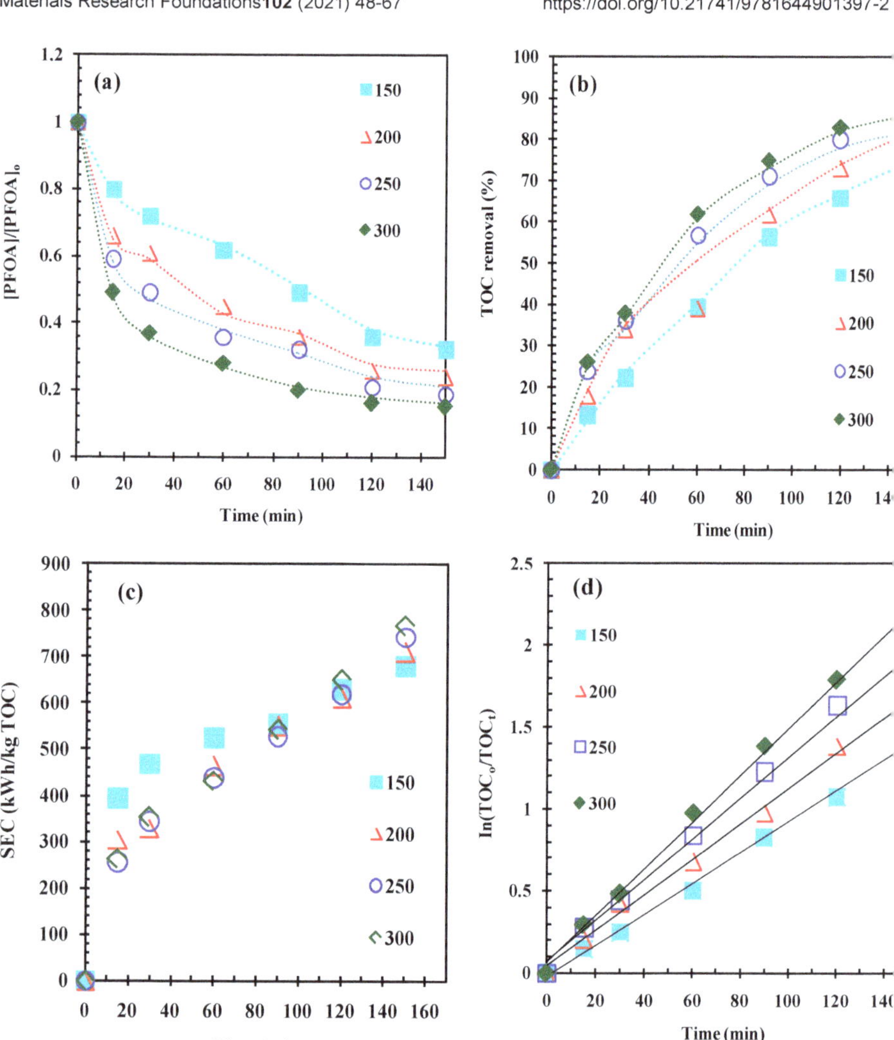

Figure 3: Effect of NaCl dose (mg L^{-1}) for the PFOA mineralization (a) PFOA removal, (b) TOC removal efficiency, (c) kinetics studies, and (d) SEC (kWh/kg TOC) at j=168.9 Am^{-2}, pH=3.8 and C = 100 mg L^{-1} at different time intervals.

Materials Research Forum LLC
https://doi.org/10.21741/9781644901397-2

3.4 Mineralization mechanism of PFOA

The EO mechanisms of organic pollutants removal concerned the OH• mediated oxidation or e⁻ transfer from target molecules to the anodic surface via the direct or mediated EO. Many previous reports have been confirmed that OH• mediated oxidation was not more effective for PFOA degradation, the EO studies of PFOA removal was mainly ascribed to the direct oxidation [1]. GC-MS analysis was employed to analyze the samples collected at different time intervals during the treatment at the optimal treatment conditions and identified the various intermediates by-products. There are different paths of PFOA degradation as describe in the plausible degradation mechanism represented in Fig. 4 based on the previously reported studies on PFOA degradation and intermediates identified [1,25]. Electrochemically generated main active species such as electrons (e^-), OH•, and $O_2^{\bullet-}$ radicals were formed during EO. At the beginning, C-F bond more preferable cleavage via the interaction of electrochemically generated e–(Eqs.15-18).

$$C_7F_{15}COO^- \rightarrow C_7F_{15}COO^\bullet + e^- \quad (15)$$

$$C_7F_{15}COOH + e^- \rightarrow {}^\bullet C_7F_{14}COOH + F^- \quad (16)$$

$${}^\bullet C_7F_{15}COOH + H_2O \rightarrow C_7F_{14}COOH + {}^\bullet OH \quad (17)$$

$$C_7F_{14}COOH + e^- \rightarrow {}^\bullet C_7F_{13}COOH + F^- \quad (18)$$

In addition, the inductive effect (IE) of the carboxyl group of PFOA supports the more easily C–F bond breaking via the attack of electrochemically generated e^- (Shown in Fig. 4).

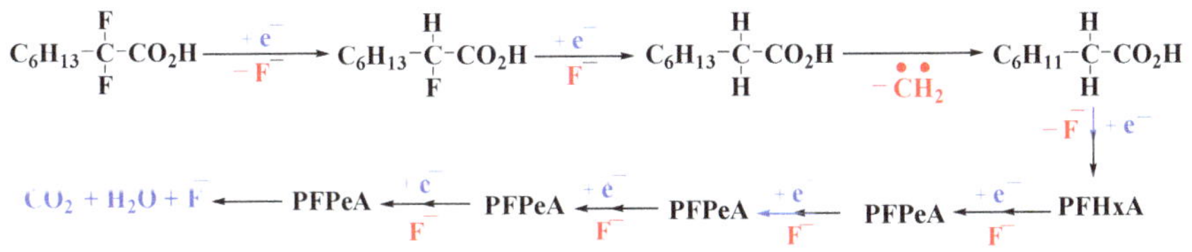

Figure 4: EO mechanism of PFOA degradation on Ti/SnO₂-Sb₂O₅/PbO₂ anode.

Materials Research Foundations **102** (2021) 48-67 — https://doi.org/10.21741/9781644901397-2

C_6F_{13} radical, carbene (:CH_2) and COOH radical were generated by the EO reactions of $C_7F_{13}H_2COOH$, then both C_6F_{13} and COOH radicals could interact to form $C_6F_{13}COOH$ (PFHpA) (Eq. 19 and 20):

$$C_7F_{13}H_2COOH \rightarrow C_7F_{13} + :CH_2 + {}^{\bullet}COOH \tag{19}$$

$${}^{\bullet}C_7F_{13} + {}^{\bullet}COOH \rightarrow C_7F_{13}COOH \tag{20}$$

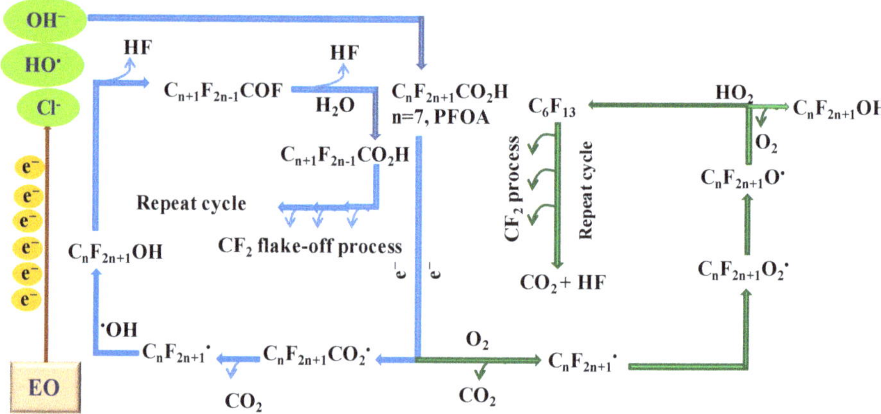

Figure 5: Plausible degradation mechanism of PFOA at the optimal treatment condition of electrochemical oxidation with Ti/SnO2-Sb2O5/PbO2 anodes.

Table 2: Different m/z values and their corresponding intermediate by-products.

m/z	412	368	364	318
Fragment ions	$C_7F_{15}CO_2^-$	$C_7F_{15}^-$	$C_6F_{13}CO_2^-$	$C_6F_{13}^-$
m/z	312	264	218	212
Fragment ions	$C_5F_{11}CO_2^-$	$C_4F_9CO_2^-$	$C_4F_9^-$	$C_3F_7CO_2^-$
m/z	168	164	118	114
Fragment ions	$C_3F_7^-$	$C_2F_5CO_2^-$	$C_2F_5^-$	$CF_3CO_2^-$

As a result, PFHpA was decaying into $C_6F_{13}CO_2^-$ (PFHxA) and $C_5F_{11}CO_2^-$ (PFPeA), which were further continually degraded into $C_3F_7CO_2^-$ (PFBA), $C_2F_5CO_2^-$ (PFPA) and $CF_3CO_2^-$ (TPA) in the similar way and was lastly, mineralized to CO_2 and F^- ions in a stepwise

 Materials Research Forum LLC
https://doi.org/10.21741/9781644901397-2

approach [26,27]. The detail of GC-MS analysis at different time intervals was used to know the decomposition products and CF_2 flake-off process summarized in Table 2. The ions peaks $C_7F_{15}CO_2$ (m/z =413), $C_7F_{17}CO_2$ (m/z = 385), C_7F_{15} ((m/z = 369) (Eq.19), and $C_6F_{13}CO_2$ (m/z = 363) (Eq.20), which can give significant information for the CF_2 mechanism [25]. Based on the literature and experimental results [19,25], a plausible degradation mechanism including radical reaction, decarboxylation, decomposition and hydrolysis was described for the PFOA removal at the Ti/SnO_2-Sb_2O_5/PbO_2 anodes (Fig. 5). The Carboxyl group (–COOH) of PFOA transfer an e^- to the anode and exist in form of PFOA radicals ($C_7F_{15}CO_2^{\bullet}$) (Eq. 15) via the path II [18]. The perfluoroalkyl radicals were formed via the decarboxylation reactions of very highly unstable $C_7F_{15}CO_2^{\bullet}$ radicals (Eq.14).

$$C_7F_{15}CO_2^{\bullet} \rightarrow C_7F_{15}^{\bullet} + CO_2 \quad (21)$$

$$C_7F_{15}COOH + O_2^{\bullet-} \rightarrow C_7F_{15} + CO_2\uparrow \quad (22)$$

Then $C_7F_{15}^{\bullet}$ react with H_2O, O_2, or $OH^{\bullet}$ to form $C_7F_{15}OH$ and $C_6F_{13}COF$ (Eq. 21–27) by the elimination of H^+ and F–[19].Moreover,$C_7F_{15}^{\bullet}$ radicals may react with O_2 (Eq. 26) and another perfluoroheptylperoxy radical ($R_fCO_2^{\bullet}$) to form perfluoroalkoxy radical ($C_7F_{13}O^{\bullet}$) via the Eq. 27

$$C_7F_{15}^{\bullet} + H_2O \rightarrow C_7F_{15}OH + H^+ \quad (23)$$

$$C_7F_{15}^{\bullet} + HO^{\bullet} \rightarrow C_7H_{15}OH \quad (24)$$

$$C_7F_{13}OF + H_2O \rightarrow C_6F_{13}CO_2^- + HF + H^+ \quad (25)$$

$$C_7F_{15}^{\bullet} + O_2 \rightarrow C_7F_{13}O_2^{\bullet} \quad (26)$$

$$C_7F_{13}O_2^{\bullet} + R_fCO_2^{\bullet} \rightarrow C_7F_{15}O^{\bullet} + R_fCO^{\bullet} + O_2 \quad (27)$$

$$C_7F_{15}O^{\bullet} \rightarrow C_7F_{13}^{\bullet} + COF_2 \quad (28)$$

$$COF_2 + H_2O \rightarrow HF + CO_2 \quad (29)$$

Subsequently, the $C_7F_{13}O^{\bullet}$ may follow the two degradation pathways: (i) hydrolysis with H_2O to form alcohol $C_7F_{15}OH$ (Eqs. 23 and 24) and (ii) decomposition toperfluorohexyl radical ($C_6F_{13}^{\bullet}$) and carbonyl fluoride (COF_2) via the Eq. 28. Finally, COF_2 will again hydrolysis with complete decay (Eq. 29).

Conclusion

Present study proposed the electro oxidative degradation of PFOA with Ti/SnO_2-Sb_2O_5/PbO_2 electrode. More than 90% PFOA and 82%TOC removal efficiency were formed at j of $168.94 Am^{-2}$, pH of 3.58 and NaCl concentration of 250 mg L^{-1} after 120 min of electrolysis with keeping the electrode gap of 1 cm fixed. The reaction kinetic of TOC reduction was followed the pseudo-first-order kinetics. The value of k for PFOA was increased with increasing the j and electrolyte concentration,whereas the removal efficiencies were decreased with increasing the solution pH. Active species such as chloro species (Cl^-, ClO^-, ClO_2^-, HClO), $OH^{\bullet}$, and $O_2^{\bullet -}$ were produced during electrolysis and play a significant role in the mineralization of PFOA. The short-chain (C_2–C_7) perfluorocarbons (PFCs) intermediate were detected and a plausible degradation mechanism of PFOA degradation was described during electrolysis based on the intermediate by-products formed during electrolysis.

References

[1] J. Niu, Y. Li, E. Shang, Z. Xu, J. Liu. Electrochemical oxidation of perfluorinated compounds in water. Chemosphere 146 (2015) 526-538.https://doi.org/10.1016/j.chemosphere.2015.11.115

[2] S. Li, G. Zhang, W. Zhang, H. Zheng, W. Zhu, N. Sun, Y. Zheng, P. Wang. Microwave enhanced Fenton-like process for degradation of perfluorooctanoic acid (PFOA) using Pb-$BiFeO_3$/rGO as heterogeneous catalyst. Chem. Eng. J. 326 (2017) 756–764.https://doi.org/10.1016/j.cej.2017.06.037

[3] S. Khan, X. He, J.A. Khan, H.M. Khan, D.L. Boccelli, D.D. Dionysiou. Kinetics and mechanism of sulfate radical- and hydroxyl radical-induced degradation of highly chlorinated pesticide lindane in UV/peroxymonosulfate system. Chem. Eng. J. 318 (2017)135–142.https://doi.org/10.1016/j.cej.2016.05.150

[4] M.-K. Kim, T. Kim, T.-K. Kim, S.-W. Joo, K.D. Zoh. Degradation mechanism of perfluorooctanoic acid (PFOA) during electrocoagulation using Fe electrode. Sep. Purif. Technol. 247 (2020) 116911. https://doi.org/10.1016/j.seppur.2020.116911

[5] M.-J. Chen, S.-L. Lo, Y.–C. Lee, C.-C. Huang. Photocatalytic decomposition of perfluorooctanoic acid by transition-metal modified titanium dioxide. J. Hazard. Mater. 288 (2015) 168–175.https://doi.org/10.1016/j.jhazmat.2015.02.004

[6] Q. Zhuo, S. Deng, B.Yang, J. Huang, G. Yu. Efficient electrochemical oxidation of perfluorooctanoate using a Ti/SnO_2-Sb-Bi anode, Environ. Sci. Technol. 45 (2011) 2973–2979.https://doi.org/10.1021/es1024542

[7] Z. Li, Q. Yang, Y. Zhong, X. Li, L. Zhou, X. Li, G. Zeng. Granular activated carbon supported iron as a heterogeneous persulfate catalyst for the pretreatment of mature landfill leachate. RSC Adv. 6 (2016) 987–994.https://doi.org/10.1039/C5RA21781D

[8] M.-J. Chen, S.-L. Lo, Y.-C. Lee, J. Kuo, C.-H. Wu. Decomposition of perfluorooctanoic acid by ultraviolet light irradiation with Pb-modified titanium dioxide. J. Hazard. Mater. 303 (2016) 111–118.https://doi.org/10.1016/j.jhazmat.2015.10.011

[9] Y.C. Lee, S.-L. Lo, P.T. Chiueh, D.G. Chang. Efficient decomposition of perfluorocarboxylic acids in aqueous solution using microwave-induced persulfate. Water Res. 43 (2009) 2811–2816.https://doi.org/10.1016/j.watres.2009.03.052

[10] L.-A.P. Thi, H.-T. Do, S.-L. Lo. Enhancing decomposition rate of perfluorooctanoic acid by carbonate radical assisted sonochemical treatment. Ultrasonics Sonochem. 21 (2014) 1875–1880.https://doi.org/10.1016/j.ultsonch.2014.03.027

[11] Y.-b. Hu, S.-L. Lo, Y.-F. Li, Y.-C. Lee, M.-J. Chen, J.-C. Lin. Autocatalytic degradation of perfluorooctanoic acid in a permanganate-ultrasonic system. Water Res. 140 (2018) 148-157.https://doi.org/10.1016/j.watres.2018.04.044

[12] V. Mulabagal, L. Liu, J. Qi, C. Wilson, J.S. Hayworth. A rapid UHPLC-MS/MS method for simultaneous quantitation of perfluoroalkyl substances (PFAS) in estuarine water. Talanta 190 (2018) 95-102.https://doi.org/10.1016/j.talanta.2018.07.053

[13] Z.W. Du, S.B. Deng, Y. Bei, Q. Huang, B. Wang, J. Huang, G. Yu. Adsorption behavior and mechanism of perfluorinated compounds on various adsorbent – a review. J. Hazard. Mater. 274 (2014)443–454.https://doi.org/10.1016/j.jhazmat.2014.04.038

[14] Y.P. Bao, J.F. Niu, Z.S. Xu, D. Gao, J.H. Shi, X.M. Sun, Q.G. Huang, Removal of perfluorooctane sulfonate (PFOS) and perfluorooctanoate (PFOA) from water by coagulation: mechanisms and influencing factors. J. Colloid Interface Sci. 434 (2014) 59–64.https://doi.org/10.1016/j.jcis.2014.07.041

[15] Y.-C. Lee, Y.-F. Li, M.-J. Chen, Y.-C. Chen, S.-L. Lo. Efficient decomposition of perfluorooctanic acid by persulfate with iron-modified activated carbon. Water Res. 174 (2020) 115618.https://doi.org/10.1016/j.watres.2020.115618

[16] B. P. Chaplin. Critical review of electrochemical advanced oxidation processes for water treatment applications. Environ. Sci.: Processes Impacts, 2014, 16, 1182. https://doi.org/10.1039/C3EM00679D

[17] W. Wu, Z.-H. Huang, T.-T. Lim. Recent development of mixed metal oxide anodes for electrochemicaloxidation of organic pollutants in water. Appl. Catal. A: General 480 (2014) 58–78.https://doi.org/10.1016/j.apcata.2014.04.035

[18] P.V. Nidheesh, M. Zhou, M.A. Oturan. An overview on the removal of synthetic dyes from water by electrochemical advanced oxidation processes. Chemosphere 197 (2018) 210-227.https://doi.org/10.1016/j.chemosphere.2017.12.195

[19] H. Lin, J. Niu, S. Ding, L. Zhang, Electrochemical degradation of perfluorooctanoic acid (PFOA) by Ti/SnO_2–Sb, Ti/SnO_2–Sb/PbO_2 and Ti/SnO_2-Sb/MnO_2 anodes, Water Res. 46 (2012) 2281-2289.https://doi.org/10.1016/j.watres.2012.01.053

[20] H. Zhao, J. Gao, G. Zhao, J. Fan, Y. Wang, Y. Wang. Fabrication of novel SnO_2-Sb/carbon aerogel electrode for ultrasonic electrochemical oxidation of perfluorooctanoate with high catalytic efficiency. Appl. Catal. B Environ. 136 (2013) 278-286.https://doi.org/10.1016/j.apcatb.2013.02.013

[21] J. Niu, H. Lin, C. Gong, X. Sun. Theoretical and experimental insights into the electrochemical mineralization mechanism of perfluorooctanoic acid. Environ. Sci. Technol. 47 (2013)14341-14349.https://doi.org/10.1021/es402987t

[22] Q. Zhuo, X. Li, F. Yan, B. Yang, S. Deng, J. Huang, G. Yu. Electrochemical oxidation of 1H, 1H, 2H, 2H-perfluorooctane sulfonic acid (6:2 FTS) on DSA electrode: operating parameters and mechanism. J. Environ. Sci. 26 (2014) 1733-1739.https://doi.org/10.1016/j.jes.2014.06.014

[23] Q. Zhuo, S. Deng, B. Yang, J. Huang, B. Wang, T. Zhang, G. Yu, Degradation of perfluorinated compounds on a boron-doped diamond electrode. Electrochim. Acta 77 (2012) 17-22.https://doi.org/10.1016/j.electacta.2012.04.145

[24] S. Singh, S.-L. Lo, V.C. Srivastava, A.D. Hiwarkar. Comparative study of electrochemical oxidation for dye degradation: Parametric optimization and mechanism identification. J. Environ. Chem. Eng. 4 (2016) 2911–2921.https://doi.org/10.1016/j.jece.2016.05.036

[25] B. Xu, J. L. Zhou, A. Altaee, M. B. Ahmed, M.A. H. Johir, J. Ren, X. Li. Improved photocatalysis of perfluorooctanoic acid in water and wastewater by Ga_2O_3/UV

system assisted by peroxymonosulfate. Chemosphere 239 (2020) 124-722.https://doi.org/10.1016/j.chemosphere.2019.124722

[26] Y.-C. Lee, S.-L. Lo, J. Kuo, C.-P. Huang, Promoted degradation of perfluorooctanic acid by persulfate when adding activated carbon, J. Hazard. Mater. 261 (2013) 463– 469.https://doi.org/10.1016/j.jhazmat.2013.07.054

[27] J.-C. Lin, C.-Y. Hu, S.-L. Lo. Effect of surfactants on the degradation o perfluorooctanoic acid (PFOA) by ultrasonic (US) treatment. UltrasonicsSonochem. 28 (2016) 130–135.https://doi.org/10.1016/j.ultsonch.2015.07.007

Materials Research Forum LLC
https://doi.org/10.21741/9781644901397-3

Chapter 3

Recent Developments of Membrane Technology for Wastewater Treatment

Mahendra S. Gaikwad *

Department of Chemical Engineering, Dharmsinh Desai University, Nadiad-387001, Gujarat, India

* mahendra14g@gmail.com, mahendragaikwad.ch@ddu.ac.in

Abstract

The worldwide one major and important issue is the increasing shortage of freshwater. Water is polluted by various category of pollutant such as heavy metal, organic toxic chemical, dyes and others. In such situation providing better solutions for water treatment is a major challenge for researchers. Various techniques have been used in wastewater treatment applications but among those techniques the membrane technology is the most promising technology. This chapter contains recent progress of membrane technology for advanced wastewater treatment, is systematically summarize. This review includes introduction about different membrane technology such as microfiltration (MF), ultrafiltration (UF), nanofiltration (NF) and reverse osmosis (RO). Current status of each membrane separation techniques, membrane cleaning techniques, challenges and promising solutions for various wastewater treatment have been discussed.

Keywords

Membrane Technology, Wastewater, Pollutant Removal, Membrane Fouling, Membrane Cleaning

Contents

Materials Research Foundations **102** (2021) 68-105 https://doi.org/10.21741/9781644901397-3

1. Introduction

The throughout worldwide contamination of ground and surface water has become a major environmental issue for the last decades. The anthropogenic activities are the main reason of water pollution. That include, various industrial waste, sewage and fertilizers. The major sources of wastewater are industrial and municipal applications [1]. The various industries include tannery [2], dye [3], petroleum [4], electroplating [5], pharmaceutical [6], chemical industries [7], agrochemical [8], pesticide [9] and many more industries producing wastewater and it is harmful to environment, human and animal life.

The various technologies already in used for wastewater treatment, including membrane filtration [10,11], adsorption [12,13], coagulation [2], chemical precipitation [14], electrocoagulation [14], ion exchange [15], capacitive deionization [1,16–17], biosorption and bioaccumulation [18], advanced oxidation process [19]. In the past few decades, membrane separation processes have received significant attention for wastewater treatment due to its high energy efficiency and low environmental impact [20].

This book chapter presents recent progress of membrane technology for advanced wastewater treatment in last three years. This chapter review includes introduction about different membrane separation process namely, microfiltration (MF), ultrafiltration (UF),

Materials Research Foundations **102** (2021) 68-105 https://doi.org/10.21741/9781644901397-3

nanofiltration (NF), and reverse osmosis (RO). The current status of each membrane separation techniques, challenges and promising solutions for various wastewater treatment have been discussed.

2. Information about different pressure driven membrane processes

2.1 MF process

MF is one of the well-known oldest pressure driven membrane process used in water treatment [21]. It has been used to remove micrometer sized matter (suspended particles, large bacteria, major pathogens, yeast cells and proteins) based on the principle of physical separation [22]. MF membranes have 0.1 –5 μm range of pore diameters. The MF membranes need low hydrostatic pressures to reject a high contaminant [23–25]. The MF process also used as a pre-treatment for RO or NF membrane processes to decrease the fouling potential [26].

2.2 UF process

The UF membrane process can remove compounds ranging between 0.05–10 μm successfully and this process is operated between MF and RO [27]. UF membranes are impressively noticeable water filters to remove macromolecules, suspended matter, pathogenic and microorganisms among others with low energy consumption [28]. However, UF membrane process has some limitations such as incompetence to reject any dissolved organic substances from the water and needs consistent cleaning to achieve high water pressure flow [29].

2.3 NF Process

NF process operated between UF and RO membranes process. The NF membrane pore size is normally of the order of 1 nm. The molecular weight cut-off (MWCO) range for NF membrane is 100–5000 Da. NF membranes work at moderate applied pressures with no phase change and it is capable of successfully removing small organic molecules and multivalent salts. The NF is a highly attractive process for selectivity and is cost effective compared to other traditional processes. Thus, the NF process has extensive applications in water and wastewater treatment [30].

2.4 RO Process

The RO membrane process is the most attractive and globally used technology for desalination and wastewater treatment. It has been used for production of clean water for different application purpose such as drinking, agriculture, and industries. The RO process efficiency depends on the operating parameters and characteristics of the feed water. In this process external pressure is applied to the feed water stream to overcome osmotic pressure [31,32]. The RO membranes have <1 nm pore size [32, 33]. The clean water is obtained by rejecting salt and other minerals when applied pressure is higher than the osmotic pressure ($\Delta p > \Delta \pi$). The salt rejection and water flux have significant role in the performance evaluation of RO membranes [31,32].

3. Current status of MF, UF, NF and RO membrane processes

3.1 MF current progress discussion

This section includes the recent progress on MF reported in the last three years. MF process are used as single as well as combined or hybrid process.

Ba and co-authors investigated the efficiency of a hybrid bioreactor (HBR) and hollow fiber MF membranes to remove a mixture of 14 PPs from municipal wastewater. The results show that complete removal of PPs from wastewater was obtained by using a HBR and MF membrane process [34]. Goswami and his team reported a study on the treatment of three different industries wastewater by biodegradation followed by a MF membrane process. This integrated system performance indicated very high COD removal. This integrated system also labels as cost-effective approach for different industrial wastewaters with vision of valuable resource recovery [35]. Jafari and co-authors developed the tubular ceramic MF membranes by using activated carbon for the treatment of oily wastewater. The outcome of the experiment indicates the permeation flux and rejection of total organic carbon improved due to existence of activated carbon and natural zeolite in the membrane structure. The TOC rejection was observed up to 99.99% with Mullite-Zeolite-Activated carbon membrane [36]. Saja and co-workers reported their work on flat ceramic MF development by using natural perlite and applied for treatment of industrial wastewater. That membrane was prepared by a uniaxial pressing process and sintering at optimized temperature. The prepared membrane shows 52.11% porosity, 1.70 μm average pore size, 21.68 MPa mechanical resistance and also it has good chemical resistance. The turbidity retention was found to be 97 and 96% for agro-food and tannery effluents respectively from experiments [37]. Pan and co-authors reported work on a novel electrochemical MF process. In that process the coal-based carbon membrane was used for the removal of bisphenol A (BPA) from wastewater. The membrane plays the roles of filter and anode simultaneously. The electrochemical MF process shows the excellent performance for BPA removal because of the synergistic effect between electrochemical oxidation and membrane filtration [38]. Beqqour et al. prepared the low-cost ceramic MF pozzolan membrane using micronized phosphate by dry uniaxial pressing process and sintering. The

Materials Research Foundations **102** (2021) 68-105
https://doi.org/10.21741/9781644901397-3

results indicate that improvement was found in mechanical strength, microstructure and permeability of the membrane on increased temperature. The experimental results showed that the membrane could almost reject all turbidity for tannery effluent and aluminum chloride suspension [39]. Mouiya et al. studied the development of MF ceramic membrane by uniaxially compressing natural Moroccan materials. The natural phosphate can be helpful to porosity development. The prepared MF membrane demonstrated excellent turbidity removal performance for synthetic solution (99.86%), effluent (99.80%), and seawater (99.62%). After the filtration experiments MF ceramic membrane show good mechanical stability and microstructure [40]. Arribas et al. reported work on modification of the structural and morphological properties of polysulfone electrospun nanofibrous membranes by heat post-treatment method to improve filtration performance. The modified membrane shows the higher PWP than commercial membranes with lower pressures requirement and thus it drops energy consumption in filtration [41]. Li and co-authors worked on CuO/carbon membrane fabrication. The electrocatalytic performance of the membrane was improved by uniformly deposition of CuO electrocatalysts on a coal-based carbon membrane. The obtained membrane showed superior removal capability and it showed great potential, high permeability and removal efficiency for the removal of small-sized organic pollutants [42]. Sheikhi et al. studied development of kaolinitic clay-based ceramic membrane for investigation of inline coagulation-MF (IMF) process performance for oily wastewater. The overall results indicated that the IMF system could be reflected as a capable system for filtration performance intensification and membrane fouling mitigation [43]. Hatimi et al. reported work on preparation of low-cost MF membrane from natural clay and pyrrhotite ash solid waste. They made changes in the surface morphology and improved the hydrophobicity of the membrane. The prepared membrane was used for the treatment of industrial wastewaters (dairy and tannery). The membrane performance indicates the almost complete turbidity, oil and fat reduction even at high flows [44]. Manni et al. reported a study on the preparation of low-cost ceramic MF membrane made from natural magnesite. The experimental results clearly demonstrated that magnesite membrane effectively clarifies textile wastewater. The result shows that 99.9 and 69.7% removal of turbidity and COD achieved respectively [45]. Zhang et al. developed diatomite hybrid MF carbon membranes (MFCMs) for removal of oil from oily wastewater. The maximum oil rejection was found to be 98.2% for 200 mg/L of oily wastewater. The permeation flux could be almost completely recovered after regeneration of the MFCMs by simple water backwashing [46]. Homem et al. modified the polyethersulfone MF membranes with polyethilenimine (PEI) and graphene oxide (GO). The best result indicates the 97.8% rejection of Blue Corazol dye and 99.4 L m^{-2} h^{-1} bar^{-1} pure water permeability. The membrane shows > 80% flux recovery ratio after cleaning [47]. Mouiya and co-authors reported clay-based ceramic membrane development. In that corresponding support was prepared using a clay/banana peels mixture followed by sintering. The MF membrane was used for the treatment of industrial wastewater. The Experimental results show high turbidity retention and dye rejection [48]. Liu et al. reported work on the fabrication of membranes from polyvinyl alcohol and polyacrylonitrile by electrospinning approach. The membrane demonstrated good adsorption capacity for Cd (II) ions. This study also found regeneration efficiency of above 90% after three recycle operations. Thus, this membrane could be reused for several times [49]. Abadikhah et al. reported the work SiO_2 nanoparticles deposition on the outer surface of Si_3N_4 hollow fiber membrane through the sol-gel process. It results found that the reduction in the surface roughness and hydrophilicity of the membrane was enhanced. Experimental outcomes show that high separation of oil performance was achieved [50]. Chen and co-authors report work on a waste-to-resource strategy for rational preparation of low cost ceramic membranes, which can be applied for the treatment of heavy metal-laden sludges and the separation of oil-in-water (O/W) emulsions. Outcomes indicate this membrane exhibited high rejection and high flux [51]. In the study reported by Plakas et al. catalytic membrane reactor consisting of a tubular porous alumina membrane with embedded iron oxide nanoparticles was used for pharmaceutical diclofenac oxidation. Significant diclofenac oxidation/mineralization was obtained by the catalytic membrane reactor [52]. The recent progress on MF performance in wastewater treatment are systematically shown in table 1.

3.2 UF current progress discussion

This section includes the recent progress on UF membrane processes reported in the last three years. Bhattacharya and co-authors developed the ceramic UF membrane by using green synthesized zinc oxide nanoparticles. The obtained UF membrane was used for atenolol and ibuprofen drugs removal from a synthetic solution. The Results showed that removal of atenolol and ibuprofen were found to be 96% and 99% respectively by single a step UF process [53]. A fungal bioactive UF membrane from filamentous fungi Aspergillus carbonarius M333 was developed by Isik and co-authors. The prepared membrane was used for textile industry wastewater treatment. The fungal bioactive ultrafiltration membrane experimental performance indicate that decolourization of wastewater and COD reduction were obtained 91% and 73.2% respectively [54]. Bhattacharya et al. developed the high flux ceramic UF membrane and it was applied to ciprofloxacin removal from synthetic feed solution. In that study the membrane was prepared with copper oxide and TiO_2 nanoparticles as composite. The obtained UF membrane performance showed effective removal of ciprofloxacin up to 99.5% [55]. Derouich et al. prepared a new composite membrane by polypyrrole layer deposition on a flat porous support based on a natural geomaterial. This obtained membrane was tested for

Materials Research Forum LLC
https://doi.org/10.21741/9781644901397-3

dye textile effluent treatment. At optimized conditions, the prepared UF membrane removed Congo red dye up to 98% [56]. Li et al. fabricated a hybrid UF membrane and that hybrid UF membrane demonstrated high water flux and outstanding nickel ion adsorption capacity was obtained in the treatment of high-salinity synthetic wastewater [57]. Shakak and co-author reported a study on the preparation of nanocomposite UF membrane from Polysulfone, Polyvinylpyrrolidone and SiO_2 using the phase inversion method. The investigation of membrane performance presented that removal performance of amoxicillin increased from 66.52% to 89.81% by increasing SiO_2 nanoparticles from 0 to 4 wt% respectively. The addition of SiO_2 nanoparticles also enhanced the membrane resistance against fouling [58]. Ye et al. fabricate the tight UF membranes from the PDA/polyethylenimine co-deposition triggered by ammonium persulfate. It was used for highly saline textile wastewater treatment.

Table 1. Recent progress on MF performance in wastewater treatment

Process	Membrane details	Wastewater type	Name pollutants	Feed Concentration	Removal % / efficiency/ output	Ref.
Hybrid hybrid bioreactor (HBR) and hollow fiber MF	Polysulfone hollow fiber MF membrane	municipal wastewater	Mixture of 14 pharmaceutical products	13 spiked pharmaceutical products at 10 mg/L and native acetaminophen at 35.5 mg/L	90%	[24]
biodegradation followed by MF membrane filtration.	low-cost ceramic membrane	A: dairy wastewater B:paper and pulp wastewater C:biomass gasifiation wastewater	COD removal	A: 2234 mg L^{-1} B: 937 mg L^{-1} C: 2157 mg L^{-1}	A:92.7%, B:87.6% C:88.2%	[35]
MF	hybrid ceramic/carbon microfiltration membrane (Mullite-Zeolite-Activated carbon membrane)	oily wastewater treatment	A: Oil B:TOC	A: 500/1000/1500 mg/L B: 1500 mg/L	A: 98.5/ 98.2/97.4% B: 99.99%	[36]
MF	ceramic membrane based on Moroccan natural perlite	agro-food (A) and tannery effluents (B)	Turbidity (NTU)	A: turbidity: 224 NTU B: turbidity: 36 NTU	A: 97% and B: 96%.	[37]
Electrochemical MF process	coal-based carbon membrane	bisphenol simulated wastewater	bisphenol A and COD	BPA (50ppm)	97% and 90%,	[38]

MF	ceramic microfiltration pozzolan membrane	aluminum chloride suspension (A) and tannery industry wastewater(B)	Turbidity	A: turbidity: 1743 NTU B: turbidity: 201NTU	A: 99% B: 97%	[39]
MF	ceramic MF membrane (based on natural clay and Moroccan phosphate)	Beamhouse effluent (A), raw seawater (B), and aluminium chloride(C)	Turbidity	A: (turbidity: 760 NTU) B: (turbidity: 120 NTU) C. (turbidity: 725 NTU)	A: (99.80%), B:(99.62%) C: (99.86%).	[40]
MF	polysulfone electrospun nanofirous membranes	Synthetic wastewater	humic acid	15 mg/L	Separation factor percentage 77.6	[41]
electrocatalytic MF	electrocatalytic microfitration CuO/Carbon membrane	Synthetic wastewater	A: Rhodamine B (RhB) B: COD C: TOC	RhB: 300 mg/L	A: 99.96%, B: 71.82% and C: 64.29%	[42]
Inline coagulation-MF (IMF) process	kaolinitic clay-Based ceramic membrane	oily wastewater	Oil	921 mg/L	98.06%	[43]
MF	natural clay and pyrrhotite ash solid waste based MF membrane	A: Tannery effluent B: Dairy effluent	A1: Turbidity A2: Conductivity A3: Cr B1:Turbidity B2:Conductivity B3:Oil and fat	A1: 595.00 NTU A2: 39.40 ms/cm A3: 48.31 mg/l B1: 789.00 NTU B2: 4.35 ms/cm B3: 150.02 mg/L	A1: 96% A2: 3% A3: 8% B1: 100 % B2: 3% B3: 94 %	[44]
MF	low-cost ceramic microfiltration membrane made from natural magnesite	Textile wastewater.	Turbidity	144.5 ± 0.6 NTU	99.9%.	[45]
MF	Diatomite hybrid microfiltration carbon membrane	oily wastewater	Oil	200 mg/L	98.2%	[46]

MF	polyethersulfone microfiltration membrane surface modified with graphene oxide	Textile industries wastewater.	Blue Corazol (BC) dye rejection	10 mg/L	97.8%	[47]
MF	ceramic membrane from clay and banana peel powder	tannery wastewater(A) textile wastewater (B)	A1:Turbidity A2: COD B1:Turbidity B2: COD	A1: 554 NTU A2: 2413 mg L^{-1} B1: 250 NTU B2: 536 mg L^{-1}	A1: 99.99% A2: 90.38% B1: 99.98% B2:85.72%	[48]
MF	Nanofibrous PAN-PVA composite membranes	Contaminated water	Nanoparticles and heavy metal ions A:Cr (VI), B:Cd (II)	A: 40-120 mg/L B: 20-60 mg/L	adsorption capacity A: 66.5 mg/(g membrane) B: of 33.6 mg/(g membrane) rejection ratios higher than 99.3%	[49]
MF	SiO_2 nanoparticles modified Si_3N_4 hollow fiber membrane	oily wastewater	Oil	1000 ppm	95%	[50]
MF	Spinel-based ceramic membranes	Oily wastewater	Oil	500 mg/L	99.5%	[51]
catalytic membrane reactor	tubular porous alumina (α-Al_2O_3) membrane	synthetic solutions	Diclofenac	200 µg/L	~ 65%	[52]

The remarkable fractionation of dye/salt mixtures are attained. These bio-inspired tight UF membranes are a good alternative for textile wastewater treatment among other possible UF practices [59]. Isawi prepared the vermiculite nanoparticles incorporated polyvinylidene fluoride flat sheet membrane. The outcomes of that study clearly revealed that the application of vermiculite nanoparticles in membranes was beneficial to the enhancement of membrane performance, antifouling and mechanical properties, compared to other membranes at the same condition [60]. Yang et al. successful prepared a novel thermally stable polyimide (PI) and it was designated as characteristic candidate due to its excellent thermal stability. This PI-based UF membrane showed a rejection of 98.65% direct red 23 dye [61]. Liu and team developed the CNT/PVDF/stainless steel mesh conductive UF membrane. The new obtained membrane showed good electrical conductivity and high mechanical strength. The results showed efficient Cr(VI) removal under the application of voltage [62]. Algamdi and co-authors prepared hybrid polyethersulfone membranes incorporated with GO via a nonsolvent induced phase separation method. The GO improved membrane antifouling capabilities and membrane reusability [63]. Saja et al. fabricated the low-cost UF bentonite membrane deposited on ceramic perlite support. The bentonite layer was made by spin-coating process followed by sintering. The result showed that 97.0 and 80.1 % rejection of Direct Red 80 and Rhodamine B respectively [64]. Arefi-Oskoui et al. reported work on modification of polyethersulfone UF membrane using ultrasonic assisted functionalized MoS_2. The obtained membrane performance evaluated against treatment of oil refinery wastewater. The optimum nanocomposite UF membrane showed maximum average surface roughness and subsequently highest efficient area for filtration than other nanocomposite membranes. The experimental result of optimum nanocomposite UF membrane show COD and the turbidity were decreased by 40.8 mg/L and 2.1 NTU for the feed wastewater of 240 mg/L and 12.2 NTU respectively [65]. Ibrahim et al., fabricated polysulfone blend UF membrane using poly[styrene-alt-(N-4-benzoylglycine-maleamic acid)] cumene terminated via the phase inversion method. The membrane indicate excellent performance against removal of heavy metal ions (Pb^{2+} and Cd^{2+}). This membrane can be reused by simple acid wash [66]. Huang et al. reported work on Al^{3+}-doped TiO_2 (AT) tight UF membrane with stable anatase phase made by a modified solgel process using butyl titanate (as the precursor) and aluminum chloride (as aluminum source). The membrane showed 96.9% rejection for Alizarin Red-S for feed concentration of 250 ppm [67]. Benkhaya and team prepared composite UF membranes from blend of polysulfone and polyetherimide with flat pozzolan support. The optimized membrane showed promising performance against rejection of 96.1% and 75.8% for acid orange 74 and methyl orange respectively [68]. Yu et al. reported work on modification of Poly(vinylidene fluoride) UF membranes with poly(vinyl pyrrolidone) to improve the antifouling properties. The pore modification and the surface modification improved the membrane flux and membrane hydrophilicity respectively. The modified membranes show improved anti-fouling abilities and such membranes are appropriate for low-level radioactive wastewater treatment [69]. Benkhaya et al. prepared new organic UF membranes based on the physical blends of Polysulfone/Polyetherimide by phase inversion process. The prepared membrane was found effective for dye rejection [70]. Ahmad et al. reported work on polyvinyl chloride (PVC) membrane for improvement of surface morphology, hydrophilicity and antifouling performance by using bentonite. The oil rejection by using prepared membrane were 97.0% and 92.5% from oily-wastewater with zero salinity and 35000 ppm salinity respectively [71]. The recent progress on UF membrane application in wastewater treatment are systematically shown in table 2.

Table 2. Recent progress on UF performance in wastewater treatment

Process	**Membrane details**	**Wastewater type**	**Name pollutants**	**Concentration of Feed**	**Removal % / efficiency/ Output**	**Ref.**
UF	ZnO UF membrane	synthetic wastewater	A: atenelol B: ibuprofen	A: 505 ppb B: 1000 ppb	A: 96% B: 98%	[53]
Biodegradation and UF	Bioactive ultrafiltration membrane	textile wastewater	A: Basic Red 18 (BR 18) dye B: Remozal Brillant blue R (RBBR) C: COD	A: 10 mg/L B: 10 mg/L C: 1372 mg/L	A: 96% B: 95% C: 73.2%	[54]
UF	CuO/TiO2 ceramic ultrafiltration membrane	synthetic solution	Ciprofloxacin	500 μg/L	99%	[55]
UF	Polypyrrole/sintered pozzolan ultrafiltration membrane	synthetic solution	Congo red dye	600 ppm	98 ± 1 %	[56]
UF	PAA/ZIF-8 /PVDF hybrid ultrafiltration	high salinity wastewater	nickel (II)	$[Ni(II)]_0$=2 mg/L, $[Na^+]_0$=15000 mg/L	maximum adsorption capacity 219.09 mg-Ni(II)/g	[57]
UF	Nanocomposite ultrafiltration membrane ($PSF/PVP/SiO_2$)	aqueous solutions	amoxicillin	30 mg/L	90%	[58]
UF	bio-inspired tight ultrafiltration membrane	highly-saline textile wastewater	A: reactive dyes (reactive blue 4) B:salts($NaCl/Na_2SO_4$ mixture)	A: 2.0 g/L B:30.0 g/L	A: > 98.2% B: > 97.0%	[59]
UF	PVDF membranes incorporated with the vermiculite NPs	Wastewater	humic acid	1 g/L	94.56%.	[60]

JF	Polyimide ultrafiltration membrane	aqueous dye solution	direct red 23 dye	100 ppm	98.65%	[61]
lectroche mical membrane ystem	CNT/PVDF/stainless steel mesh conductive Ultrafiltration membrane	aqueous solution	hexavalent chromium	1 mg/L	95.2%	[62]
JF	Graphene oxide incorporated polyethersulfone hybrid ultrafiltration membranes	aqueous solution	humic acid	10 –100 ppm	94.5%	[63]
JF	Ultrafiltration bentonite membrane	aqueous solution	Direct Red 80 and Rhodamine B	100 ppm	97.0 and 80.1 %	[64]
JF	MoS_2 /polyethersulfone nanocomposite membrane	oil refinery wastewater	A: COD B: turbidity	A: 240 mg/L B:12.2 NTU	A: 40.8 mg/L (83 %) B: 2.1 NTU	[65]
UF	Polysulfone blend ultrafiltration membrane	Wastewater	heavy metal ions (A: Pb^{2+} and B: Cd^{2+})	10 ppm	A: 91.5%, B: 72.3%	[66]
UF	Al^{3+}-doped TiO_2 tight ultrafiltration membrane	aqueous solutions	Alizarin Red-S	250 ppm	96.9%	[67]
UF	Polysulfone (PSf) and polyetherimide (PEI) composite ultrafiltration membrane	aqueous dye solutions	A: acid orange 74 (AO74) B: methyl orange (MO)	50 ppm	A: 96.1 B: 75.8%	[68]
UF	Poly(vinyl pyrrolidone) modified poly(vinylidene fluoride) ultrafiltration membrane	radioactive wastewater	A: nuclide ions and B: SDBS (anionic surfactant sodium dodecyl benzene sulfonate)	A: 0.1 mg/L B: 200 mg/L	A: 98% of divalent ions Sr(II) and Co(II) B: SDBS were lower.	[69]
UF	Polysulfone/Polyetherimide	aqueous dye solutions	A: acid orange 74 (AO-74) and	50 ppm	A: 68.5% B: 64.7%	[70]

Materials Research Forum LLC
https://doi.org/10.21741/9781644901397-3

	blends organic ultrafiltration membrane		B: methyl orange (MO)			
UF	mixed-matrixpoly(vinyl chloride)-bentonite ultrafiltration membrane	saline oily wastewater	A: Oil B: salt	A: 200 ppm B: 35000 ppm	A: 97.0% B: negligible salinity	[71]
UF	polyethersulfone (PES) and regenerated cellulose (RC) commercial ultrafiltration membranes	poultry processing wastewater (Bird washer[BWW] water, Chiller water[CW])	A: biochemical oxygen demand (BOD), B:chemical oxygen demands (COD), C: total suspended solids (TSS), D: fat, oil and grease (FOG)	A: 400 mg/L B: 1000 mg/L C: 300 mg/L D: 200 mg/L	A: (up to 93%), B: (up to 94%), C: (up to 100%) D: (up to 100%) for both wastewater streams.	[72]
complexation- UF process	composite membrane (made of polysulfone and nanoclay)	aqueous solutions	anionic dye (Acid-Orange 7, AO7)	83 mg/L	99.62%	[73]

3.3 NF current progress discussion

This section includes recent progress on NF membrane processes reported in last three years. Zhao et al. fabricated a loose layered double hydroxides /polymer hybrid layer NF membrane. It tested for the desalination of textile wastewater. The membrane performance shows a sufficient permeability, high rejection of dyes and low salt rejection [74]. Zhao and co-authors prepared a loose NF composite membrane and it was used for treating textile wastewater. The membrane prepared at optimized condition show high permeability and high rejection of methyl blue dye. It also shows excellent antibacterial properties [75]. Zhang et al. prepared a photo-assisted NF membrane build from g-C_3N_4, TiO_2, CNT and GO. Photo-assisted NF membranes show a higher water flux and high methyl orange dye rejection (~100%). The NF membrane combined with photocatalysis shows the efficient removal of bisphenol A, ammonia and antibiotic from water [76]. Maryam et al. tested two types of loose NF membranes for the filtration of selected small molecular weight drugs from synthetic wastewater. The pH effect study show that behaviour of drugs changed with ph. The results showed remarkable treatment of drugs in the order of, DIC > IBU > PARA along with TOC and COD [77]. Tsehaye et al. reported a study on the combination of TiO_2 nanoparticles via blending in order to fabricate PES (TiO_2) membranes. For the reference

Materials Research Foundations **102** (2021) 68-105 https://doi.org/10.21741/9781644901397-3

purpose they used commercial PES NF membranes and Pristine. The results indicate a high water permeability and 94.9% direct red 23 rejection with obtained membrane. Therefore, PES (TiO_2) membrane a capable candidate in industrial wastewater treatment [78]. Kusworo et al. reported work on polymeric membrane modification using TiO_2 nanoparticle to improve its separation and mechanical properties. The prepared membrane was used for treatment of natural-rubber wastewater.

The addition of TiO_2 improved the hydrophilicity and mechanical strength of the membrane. The experimental results show best pollutant removal efficiency for TDS, COD, NH3, and turbidity rejections with PSf-TiO_2 membrane [79]. Guo and team prepared composite NF membranes with the layer by layer approach using plant polyphenol tannic acid with rich catechol groups and hydrophilic Jeffamine having amino groups. The obtained membrane showed high pure water permeance of 37 $Lm^{-2}h^{-1}bar^{-1}$ and also maintaining rejections higher than 90% for various dyes of molecular weight range of 269 to 1017 g mol^{-1} [80]. Bandehali et al. developed PEI based NF membranes by incorporation [L-cysteine functionalized POSS nanoparticles]. The prepared membrane demonstrates smoother surface and excellent antifouling properties than pristine PEI membrane [81]. Wang and co-authors fabricated graphene oxide/attapulgite composite membranes for the removal of dyes from wastewater. This composite of membranes showed rough hierarchical microstructure and higher surface hydrophilicity. The obtained composite membrane showed high rejection close to 100% for 7.5 mg L^{-1} RhB wastewater [82]. PEM-based NF membranes were prepared from poly(allylamine hydrochloride) and poly(acrylic acid) on polyacrylonitrile UF support. This membrane was used for rejection of Diclofenac, Naproxen, 4n-Nonylphenol and Ibuprofen from synthetic secondary treated wastewater [83]. de Souza et al. investigated the performance of two NF membranes for norfloxacin removal from a synthetic pharmaceutical wastewater. Both membranes showed high selectivity to norfloxacin and rejection of the antibiotic persisted between 87 and 99.5% [84]. Wang and team prepared GO-based membranes by grafting zwitterionic polymer. The membrane's antifouling ability improved due to the grafted zwitterionic polymer. The prepared membrane presented high rejection for different dyes namely Congo Red, Orange G and Methyl Orange [85]. Sangeetha and co-authors prepared hybrid NF membrane from chitosan/polyvinyl alcohol and montmorillonite clay. The obtained hybrid NF membrane show excellent overall performance [86]. Qin et al. reported work on preparation of 8 mol% yttria-stabilized ZrO_2 (8YSZ) NF membranes using size-controlled spherical ZrO2 nanoparticles. This prepared membrane was tested in the treatment of pesticide wastewater. The experimental performance of 8YSZ NF membranes showed the removal rate of carbofuran higher than 82% and whereas the extreme removal rate could reach 89% [87]. Cao et al. investigated separation performances of the dye, the intermediate (H-acid) and salts (NaCl) by two fabricated NF membranes having different charges and pore structural properties. Both membranes showed long-term stability and rejection of RB-5 was higher than 95%. [88]. Tavangar and co-authors prepared loose NF nanocomposite membranes from Poly(ether sulfone) by incorporating nanoparticles of cerium oxide. The membrane was prepared via a non-solvent induced phase separation approach. The obtained membranes performances were investigated by treatment of aqueous dye solutions and real textile wastewater. The nanocomposites membrane demonstrated notable better antifouling properties than the pristine membrane [89]. The various studies on NF membrane used in wastewater treatment are systematically shown in table 3.

Table 3. Recent progress on NF performance in wastewater treatment

Process	Membrane details	Wastewater type	Name pollutants	Feed Concentration	Removal % / efficiency/ Output	Re
NF	loose layered double hydroxides (LDHs) LDHs/polymer hybrid NF membrane.	simulated dye wastewater	A: Dyes (Methyl blue, acid fuchsin) B:salt (NaCl)	A: 500 ppm B: 1 g/L	A: 97.9% and 97.5% B: lower salt rejection (<3%).	[7
NF	PEGylation of plant polyphenols/polypeptide-mediated loose NF membrane	A: aqueous solutions B: Actual textile wastewater	A1: Dyes (methyl blue) A2: salts(NaCl) B1:COD, B2:NH3-N	A1:100~500 ppm A2:2–10 g/L B1: 459 mg/L B2: 8.95 mg/L	A1: 98.9% A2: 17% B1:89.1% B2:84.5%	[7
photo-assisted NF	multifunctional graphene-based NF membrane	wastewater	A: salt ions B: antibiotic (sulfamethoxazole). C: bisphenol A D: TOC E: Methyl Orange	A: 2 g/L B: 5 mg/L C: 5 mg/L D:41.33 ±0.13 mg/L	A: 67% (Na_2SO_4) B: 80% C: 82% D: 78% E: ~ 100%	[7
NF	HYDRACoRe 50 (NF50) Hydranautics (Oceanside, CA, USA)	wastewater	Pharmaceutically Active Compounds (PhACs), A:Diclofenac (DIC), B:Ibuprofen (IBU) and C:Paracetamol (PARA), D:TOC E: COD	A: 0.1g/L B: 0.1g/L C: 0.1g/L	A:99.7% B:81.2% C:49% D:95.3% E:84%	[7

Materials Research Foundations **102** (2021) 68-105
https://doi.org/10.21741/9781644901397-3

NF	Polyethersulfone (TiO2) composite membranes	Aqueous solution	Direct Red 23	0.5 g/L	94.9%	[78]
NF	nano-TiO2 loading in polysulfone membranes	Natural-rubber wastewater treatment.	A:TDS, B:COD, C:N-NH3, D:Turbidity	A: 297 mg/L B: 273 mg/L C: 18.68 mg/L D: 32 NTU	A: 14.03 % B:87.88 % C: 88.79 % D: 99 %	[79]
NF	Tannic acid and Jeffamine /polyacrylonitrile composite NF membranes	Dyes solution	A:Methyl orange (MO) B:Rose bengal sodium salt (RB) C:Methylene blue (MB) D:Congo red (CR) E:Rhodamine B (RhB)	35 μM dye solution	A: 91% B:99.5% C:98% D:99.5% E:99%	[80]
NF	[L-cysteine-co-POSS/PEI] membrane.	Aqueous solutions	A: Na_2SO_4 B: $Cr(SO_4)_2$	A: 1100 mg/L B: 500 mg/L	A: 80 % B:79%	[81]
NF	Graphene oxide/attapulgite (GO/APT) composite membranes	dyes solution	rhodamine B	7.5 mg L− 1	100%	[82]
NF	Polyelectrolyte multilayer (PEM)-based nanofiltration (NF) membranes	synthetic secondary treated wastewater	Micropollutants A:Diclofenac B:Naproxen C:Ibuprofen D: 4n-Nonylphenol	A: 0.5 μg/L B: 2.5 μg/L C: 40 μg/L D: 7 μg/L	A: 81.5% B: 66.6% C:51.6% D:61.7%	[83]
NF	NF 270 and NF 90 (Filmtec – Minneapolis, MN)	pharmaceutical effluent	Norfloxacin	50 mg L^{-1}	87 –99.5%	[84]
NF	zwitterionic polymer modifi ed GO@PDA/PES membranes	organic dye solutions	A:Congo Red, B:Orange G C:Methyl Orange	100 mg/L	A:100%, B:82% C:67%	[85]
NF	Chitosan/polyvinl alcohol/montmorillonite clay membrane	synthetic wastewater	Chromium	50 ppm	84 –88.34%	[86]
NF	yttria-stabilized ZrO2 nanofiltration membrane	synthetic pesticide wastewater	Pesticide (Carbofuran)	40/80/120/160/200 ppm	82/84/85/87/89 %	[87]
NF	M-PIP and M-PEI	complex dye-wastewater	A:reactive black 5 (RB-5) B:H-acid	A: 1 g/L B: 0.2 g /L	A: > 95% (both membrane)	[88]

					B: 96.0% (M-PEI)	
NF	cerium oxide-loaded loose nanofitration polyethersulfone membranes	Real textile wastewater.	A:Color, B:COD, C:Turbidity, D:TDS, E:TSS etc.	A: >1100 (Pt-Co) B: 3460 ± 20 mg/L C: 1340 ± 10 NTU D: 6300 ± 10 mg/L E: 250 ± 5 mg/L	A: 374 ± 20 (Pt-Co) B: 710 ± 12 mg/L C: 12.22±1.22 NTU D: 5950 ± 10 mg/L E:18.0 ± 0.5 mg/L	
NF	cerium oxide-loaded loose nanofitration polyethersulfone membranes	aqueous dye solutions	A:Congo Red B:Direct Red 23 C:Direct Red 243	100 ppm	A: 99.36% B: 99.53% C: 99.78%	[
NF	polyamide composite membranes (NE40, NE70) nonpolyamide composite membranes (MPS-34, MPS-36) Polyethersulfone (PES) based membrane (NP010, NP030)	copper-refining sulfuric acid wastewater	metal ions	More than 2000 mg/L of metal ions such as calcium, copper, iron, and magnesium. (Ca: 74 ppm, Cu:1758 ppm, Fe:429 ppm, Mg:125)	greater than 90% metal ion rejection with NE40, NE70, MPS-34	[

3.4 RO current progress discussion

The recent efforts taken by researchers on progress on RO membrane processes are summarized as follow. Pang and Zhang prepared thin film nanocomposite RO (TFN RO) membranes using hydrophilic nanoparticles. The experimental results show salt rejection was improved and it increased up to 98.6% with fluorinated silica nanoparticles at optimal 0.12% (w/v) [91]. Fujioka et al. study evaluated the variable rejection of a N-nitrosodimethylamine (NDMA), produced by fouling in the RO membrane. The fouling substances characterization in the NF treated wastewater found that fouling matters included tryptophan or tryptophan-like matters can increase NDMA rejection [92]. Ghaseminezhad et al. prepared nanocomposite RO membrane from cellulose acetate and graphene oxide with polyester fabric support. The results presented that the graphene oxide concentrations would affect the membrane's porosity, morphology, mechanical strength, hydrophilicity, roughness, and thermal stability. CA membrane modified with 1 wt% GO reject salt from synthetic seawater was 1.8 times greater compare to pristine CA membrane at 25 bar [93]. Fujioka et al., modified three commercial and one prototype RO membrane using heat treatment to achieve over 90% removal of N-nitrosodimethylamine (NDMA).The experimental outcomes show the use of heat treatment to a prototype membrane lead to high removal of 92% (1.1-log) of NDMA [94]. Sahinkaya et al., studied a pilot scale RO process (using a spiral wound membrane element) with three different operational modes to collect more information. The result showed that the permeate (conductivity 150 μS/cm) produced in the RO process can be recycled in the dyeing process as the feed conductivity was around 5500 μS/cm, at 70% water recovery [95]. Qi and co-authors fabricated the polymersomes-based RO membranes which offered excellent water permeability and water/salt selectivity mainly under feed conditions with high salt concentrations.

This study provides a standard shift in developing new-generation RO membranes for energy and cost-effective desalination processes [96]. Lopera et al. reported a study on the removal of theophylline, amoxicillin, caffeine, theobromine, and penicillin G from WWTP wastewater by RO membrane (BW30-2540). This method attained 100% contaminants removal [97]. Bastos et al., reported work evaluates the rejection of phenols from refinery wastewater by RO. The result show rejections of up to 98% of phenols and 99% of both COD and TOC by RO membrane [98]. The various studies on RO membrane used in wastewater treatment are systematically shown in table 4.

Table 4. Recent progress on RO performance in wastewater treatment.

Process	**Membrane details**	**Wastewater type**	**Name pollutants**	**Feed Concentration**	**Removal % / efficiency/ Output**	**Ref.**
RO	fluorinated silica-polyamide thin film nanocomposite RO membranes	Aqueous solution	NaCl rejection	2000 ppm	98.6%	[91]
RO	ESPA2 Nitto/Hydranautics (Osaka, Japan) and BW30, Dow Chemical Company (Midland, MI, USA),	Secondary wastewater effluent of a municipal wastewater treatment plant	N-nitrosodimethylamine (NDMA),	500 ng/L	< 70%	[92]
RO	Cellulose acetate (CA)/Graphene oxide (GO)	synthetic seawater	NaCl	25000 ppm	90	[93]

	nanocomposite membrane					
RO	heat-treated RO membrane	Aqueous solution	N-nitrosodimethylamine (NDMA)	1000 ng/L	92%	[
RO	RO pilot scale unit	biologically-treated textile wastewater	conductivity	5500 μS/cm	Less than 150 μS/cm and 70% water recovery	[
RO	Polymersomes-based RO membrane	Aqueous solution	NaCl	32000 ppm NaCl	98.9%	[
RO	DOW FILMTEC BW30-2540 membrane	Aqueous solution	emerging contaminants (including stimulants and antibiotics)	approximately 1 mg/L for each pollutant	100%	[
RO	RO membranes, BW30 and SW30	refinery wastewater	A: Phenol B: COD C:TOC	A: 128 ppm B:1605 mgO_2/L C: 399.4 ppm	A: 98% B: 99% C:99%	[
RO	tangential-flow RO bench-scale membrane plant	olive-oil washing wastewater	COD	188.7 ± 37.9 mg/L	1.5 mg/L	[

4. Challenges and promising solutions in MF, UF, NF and RO membrane processes

The third major research area in MF is membrane fouling. The unwanted deposits create the problems of pore blockage which decrease membrane permeability and selectivity [22]. Majority of the fouling studies involved MF membrane modifications. Gu et al. studied graphene oxide quantum dots (GOQDs) effects on the surface properties of ceramic membranes. The result of GOQDs modified ceramic membranes indicate improved hydrophilicity, water permeability with remain a porous structure and reduced membranes resistance [100]. Ghalamchi and co-authors developed the cost-effective blended MF membranes ZnAlCu-NLDH/polyethersulfone (PES) and gC3N4/PES nanocomposite membranes. These two membranes showed significant flux recovery ratio than the pristine PES membrane [101]. Kose-Mutlu and team modified the polymeric MF membranes by adsorption bismuth chelate (BisBAL), which have a high anti-bacterial effect on various microorganisms, on the membrane surface. The modified membranes showed strong resistance to biofouling [102]. Electric membrane bioreactor has significantly attracted interest for wastewater treatment as well as in water purification due to its outstanding performance in fouling mitigation. The application of ionic liquids in pyrrole oxidative polymerization deals a novel and effective strategy in conductive membrane preparation and helps as an excellent scheme in fouling mitigation in membrane bioreactor [103]. Gu et al. modified membrane surface with alumina presented significant changed in better antifouling properties than in pristine membranes [104]. Zhang et al. developed silver-decorated biomimetic silica nanopollens (SNPs) to modify a polyvinylidene fluoride MF membrane. The modified membrane confirmed compelling antibiofouling performance than the pristine membrane [105]. Zhang and co-authors fabricated a highly antibiofouling membrane. The excellent biofouling resistance was obtained by grafting quaternary ammonium compounds (QAC) onto a silica-decorated membrane using alkoxysilane polycondensation reaction. This membrane showed comparable water permeability, surface roughness, hydrophilicity with the pristine membrane [106].

The fouling remains a major barrier limiting UF applications in treating various waters. Thus, research on UF membrane modification is a growing sustained area of attention. Liu and co-authors prepared PVDF/PFSA-g-GO UF membranes. The prepared membranes (PVDF/PFSA-g-GO) showed greater permeability and antifouling capacity than PVDF membrane [107]. He and co-authors prepared γ-alumina double-layered UF membranes and single-layered UF membranes. The results show both single- and double-layered UF membranes exhibited comparable performance in particle size retention. The double-layered membrane presented better water flux, resistance and anti-fouling properties than the single-layered UF membrane [108]. Zhang and co-authors prepared a supported Ti^{3+} self-doped anatase TiO_2 UF membrane and its photocatalytically enhanced anti-fouling in surface water treatment [109]. Huang et al. prepared UF membranes using magnetic nanoparticles for emulsion treatment. The application of magnetic nanoparticles improved ultrafiltration performance and flux significantly [110] The biofouling is also a critical issue in the UF membrane technology. To resolve the complex biofouling problem various useful studies were reported in recent years. Khoerunnisa and co-worker reported on the preparation of chitosan/ polyethylene glycol / multi-walled carbon nanotubes / Iodine composite membranes via phase inversion technique. This composite membrane exhibited superior antibacterial activity against Escherichia coli (E. coli) and Staphylococcus aureus (S. aureus). Also, significant changes were observed in membranes properties such as the improvement of membrane hydrophilicity, porosity and mechanical strength [111]. Ahmad et al. prepared amine-functionalized zeolitic imidazole framework -8-decorated GO for UF hollow fibre membranes via the phase inversion method. These prepared membranes showed remarkable antifouling properties with a greater flux recovery ratio than the unmodified PES membranes. It also showed antibacterial competence against Escherichia coli and Staphylococcus aureus [112]. Sathya and Nithya prepared a polyetherimide

Materials Research Foundations **102** (2021) 68-105
https://doi.org/10.21741/9781644901397-3

composite membrane with tungsten oxide as anti-bacterial agent and pectin as blending additive. The result show that biofouling properties were significantly improved by the addition of metal oxide [113]. Zhang et al. prepared the mixed matrix UF membranes from guanidyl functionalized graphene/polysulfone. This membrane showed superior permeability and prominent antifouling properties, and excellent antimicrobial activity [114].

Research on NF membrane development for improving antifouling properties are the global attention of researcher. Pandey et al. developed fouling-resistant mixed-matrix NF membrane from $Ti_3C_2T_X$ (MXene)/cellulose acetate (MXene@CA) composite. The optimal membrane result indicated a significantly improved hydrophilicity and good antifouling properties [115]. Mi et al. developed effective antifouling and dye desalting NF membrane from zwitterionic functionalized monomer. The membrane exposed excellent antifouling performance with a good flux recovery ratio [116]. Koulivand and co-workers developed a novel mixed matrix NF membrane by using carbon dots blended with a polyethersulfone (PES) matrix. The PES/CDs membrane also presented significant antifouling properties. The carbon dots could be superb membrane modifier for promising wastewater treatment [117].

Shang et al. fabricated a nanostructured NF membrane with using sodium bicarbonate on a porous substrate. This obtained membrane improved fouling resistance was a result of the improved ridge-and valley nanostructure [118]. Vatanpour and Haghighat prepared the polyvinyl chloride NF membranes from Multiwalled carbon nanotubes modified with triethylenetetramine via the phase separation method and immersion precipitation technique. The result indicated the significant reduction in the rate of membrane fouling with prepared membrane [119]. Zhu et al., fabricate a thin-film composite (TFC) NF membrane with incorporation of calcium bicarbonate. The prepared membrane demonstrated outstanding antifouling performance and satisfactory flux recovery after physical cleaning for the optimized membrane [120].

Biofouling is a serious difficulty in RO membrane processes. Li and co-authors prepared thin film nanocomposite RO membranes from GOQD and silver phosphate (AP). GOQD as ideal nanofiller to enhance membrane permeability because its rich hydrophilic groups and AP has higher bactericidal effect than silver particle usually have in membrane applications. The prepared membrane demonstrated strong bactericidal property, good stability and outstanding antifouling performance [121]. The modified a polyamide RO membrane using a polyampholyte composed of cationic [2-(acryloyloxy) ethyl] trimethyl ammonium chloride (TMA) and anionic 2-carboxyethyl acrylate (CAA) via surface-initiated atom transfer radical polymerization. The optimized RO membrane showed an excellent anti-biofouling performance [122].

Materials Research Forum LLC
https://doi.org/10.21741/9781644901397-3

5. Membrane cleaning techniques and cleaning agents

The various techniques used in membrane cleaning applications are chemical, mechanical, electrical hydraulic, and ultrasonic methods [123–126]. Each of these techniques has advantages and limitations, and should be useful under appropriate conditions [127]. The chemical cleaning method is the most effective approach to improve permeability of membrane and membrane cleaning [128]. Various chemical cleaning agents includes Caustic soda (e.g. sodium hydroxide), Oxidants (e.g. hydrogen peroxide, sodium hypochlorite), acids (e.g. hydrochloric acid, sulphuric acid, citric acid) are used to remove foulants from the membrane surface and to improve the membrane efficiency [129–132].

Conclusions

Membrane separation technology is a capable technology in the field of wastewater treatment. This chapter provides systematic information of recent progress in membrane technology for wastewater treatment for the last three years. This comparative and details review provides the application of different membrane process like MF, UF, NF and RO can be applied in effectively industrial wastewater treatment. The review also includes the problems and challenges associated with the process such as membrane fouling, sensitivity, biofouling and flux recovery which are the main limitation of the current membrane technology. Researchers from worldwide are taking continuing efforts to solve these issues. From the literature, the different materials incorporation in membranes such as in graphene oxide quantum dots, carbon dots, metal oxides, nanomaterials, nanocomposite, calcium bicarbonate, sodium bicarbonate, zwitterionic functionalized monomer, among others were found effective for improved membrane performances. Thus, membrane technology will remain the most effective method for wastewater treatment.

Acknowledgments

The author is grateful to the Dharmsinh Desai University, Nadiad for providing the necessary facilities.

Materials Research Forum LLC
https://doi.org/10.21741/9781644901397-3

References

[1] A. Kalfa, B. Shapira, A. Shopin, I. Cohen, E. Avraham, D. Aurbach, Capacitive deionization for wastewater treatment: Opportunities and challenges, Chemosphere 241 (2020) 125003. https://doi.org/10.1016/j.chemosphere.2019.125003

[2] Z. Song, C.J. Williams, R.G.J. Edyvean, Treatment of tannery wastewater by chemical coagulation, Desalination 164 (2004) 249-259. https://doi.org/10.1016/S0011-9164(04)00193-6

[3] S.N. Jain, Z. Shaikh, V.S.Mane, S. Vishnoi, V.N. Mawal, O.R.Patel, P.S. Bhandari, M.S. Gaikwad, Nonlinear regression approach for acid dye remediation using activated adsorbent: Kinetic, isotherm, thermodynamic and reusability studies, Microchem. J. 148 (2019) 605-615. https://doi.org/10.1016/j.microc.2019.05.024

[4] A.S. Costa, L.P.C. Romão, B.R. Araújo, S.C.O. Lucas, S.T.A. Maciel, A. Wisniewski Jr, M.D.R. Alexandre, Environmental strategies to remove volatile aromatic fractions (BTEX) from petroleum industry wastewater using biomass, Bioresour. Technol. 105 (2012) 31-39. https://doi.org/10.1016/j.biortech.2011.11.096

[5] D. Chen, C. Zhang, H. Rong, M. Zhao, S. Gou, Treatment of electroplating wastewater using the freezing method, Sep. Purif. Technol. 234 (2020) 116043. https://doi.org/10.1016/j.seppur.2019.116043

[6] C. Gadipelly, A. Pérez-González, G.D. Yadav, I. Ortiz, R. Ibáñez, V.K. Rathod, K.V. Marathe, Pharmaceutical industry wastewater: review of the technologies for water treatment and reuse, Ind. Eng. Chem. Res. 53(2014) 11571-11592. https://doi.org/10.1021/ie501210j

[7] F.A. Nasr, H.S. Doma, H.S. Abdel-Halim, S.A. El-Shafai, Chemical industry wastewater treatment, The Environmentalist 27 (2007) 275-286. https://doi.org/10.1007/s10669-007-9004-0

[8] M.I. Pariente, J.A. Siles, R. Molina, J.A. Botas, J.A. Melero, F. Martinez, Treatment of an agrochemical wastewater by integration of heterogeneous catalytic wet hydrogen peroxide oxidation and rotating biological contactors, Chem. Eng. J. 226 (2013) 409-415. https://doi.org/10.1016/j.cej.2013.04.081

[9] H.F. Shaalan, M.Y. Ghaly, J.Y. Farah, Techno economic evaluation for the treatment of pesticide industry effluents using membrane schemes, Desalination 204 (2007) 265-276. https://doi.org/10.1016/j.desal.2006.04.032

[10] Z.V.P. Murthy, M.S. Gaikwad, Separation of praseodymium (III) from aqueous solutions by nanofiltration. Can. Metall. Q. 52 (2013) 18-22. https://doi.org/10.1179/1879139512Y.0000000042

[11] B. Van der Bruggen, C. Vandecasteele, T. Van Gestel, W. Doyen, R. Leysen, A review of pressure-driven membrane processes in wastewater treatment and drinking water production. Environ. Prog. 22 (2003) 46-56. https://doi.org/10.1002/ep.670220116

[12] M.S. Gaikwad, A.R. Deshmukh, S. Saudagar, V. Kulkarni, Synthesis and characterization of CS/MWCNTs/ES composites and its performance in removal of Cu (II) from aqueous solution, Inorg. Nano-Met. Chem. 47 (2017) 568-575. https://doi.org/10.1080/15533174.2016.1186090

[13] J.R. de Andrade, M.F. Oliveira, M.G. da Silva, M.G. Vieira, Adsorption of pharmaceuticals from water and wastewater using nonconventional low-cost materials: a review, Ind. Eng. Chem. Res. 57 (2018) 3103-3127. https://doi.org/10.1021/acs.iecr.7b05137

[14] M.S. Oncel, A. Muhcu, E. Demirbas, M., Kobya, A comparative study of chemical precipitation and electrocoagulation for treatment of coal acid drainage wastewater, J. Environ. Chem. Eng. 1 (2013) 989-995. https://doi.org/10.1016/j.jece.2013.08.008

[15] T.C. Jorgensen, L.R. Weatherley, Ammonia removal from wastewater by ion exchange in the presence of organic contaminants, Water Res.37 (2003) 1723-1728. https://doi.org/10.1016/S0043-1354(02)00571-7

[16] M.S. Gaikwad, C. Balomajumder, A.K., Tiwari, Acid treated RHWBAC electrode performance for Cr (VI) removal by capacitive deionization and CFD analysis study, Chemosphere 254 (2020)126781. https://doi.org/10.1016/j.chemosphere.2020.126781

[17] M.S. Gaikwad, C. Balomajumder, Removal of Cr (VI) and fluoride by membrane capacitive deionization with nanoporous and microporous Limonia acidissima (wood apple) shell activated carbon electrode, Sep. Purif. Technol. 195 (2018) 305-313. https://doi.org/10.1016/j.seppur.2017.12.006

[18] N. Singh, C. Balomajumder, Simultaneous biosorption and bioaccumulation of phenol and cyanide using coconut shell activated carbon immobilized Pseudomonas putida (MTCC 1194), J. Environ. Chem. Eng. 4 (2016) 1604-1614. https://doi.org/10.1016/j.jece.2016.02.011

[19] R. Mudliar, S.S. Umare, D.S. Ramteke, S.R. Wate, Energy efficient—Advanced oxidation process for treatment of cyanide containing automobile industry wastewater, J. Hazard. Mater. 164 (2009) 1474-1479. https://doi.org/10.1016/j.jhazmat.2008.09.118

[20] X. Li, Y. Mo, W. Qing, S. Shao, C.Y. Tang, J. Li, Membrane-based technologies for lithium recovery from water lithium resources: A review, J. Memb. Sci. 591 (2019) 117317. https://doi.org/10.1016/j.memsci.2019.117317

[21] W. Eykamp, Chapter 1. Microfiltration and ultrafiltration, in: R.D. Noble, S.A. Stern (Eds.), Membrane Science and Technology, Elsevier, 1995, pp. 1 –43. https://doi.org/10.1016/S0927-5193(06)80003-3

[22] S.F. Anis, R. Hashaikeh, N. Hilal, Microfiltration membrane processes: A review of research trends over the past decade, J. Water Process Eng. 32 (2019) 100941. https://doi.org/10.1016/j.jwpe.2019.100941

[23] A. Ismail, P. Goh, Microfiltration membrane, Encycl. Polymer. Nanomater. (2015) 1250 –1255. https://doi.org/10.1007/978-3-642-29648-2_159

[24] X. Qu, P. Alvarez, J.R. Werber, A. Deshmukh, M. Elimelech, The critical need for increased selectivity, not increased water permeability, for desalination membranes. Environ. Sci. Technol. Lett. 3 (2016) 112-120. https://doi.org/10.1021/acs.estlett.6b00050

[25] Q. Xiaolei, P.J.J. Alvarez, Q. Li, Applications of nanotechnology in water and wastewater treatment, Water Res. 47 (2013) 3931-3946. https://doi.org/10.1016/j.watres.2012.09.058

[26] G.K. Pearce, UF/MF pre-treatment to RO in seawater and wastewater reuse applications: a comparison of energy costs, Desalination 222 (2008) 66-73. https://doi.org/10.1016/j.desal.2007.05.029

[27] F. Qu, H. Liang, J. Zhou, J. Nan, S. Shao, J. Zhang, G. Li, Ultrafiltration membrane fouling caused by extracellular organic matter (EOM) from Microcystis aeruginosa: Effects of membrane pore size and surface hydrophobicity, J. Memb. Sci., 449 (2014) 58-66. https://doi.org/10.1016/j.memsci.2013.07.070

[28] R. Krüger, D. Vial, D. Arifin, M.Weber, M. Heijnen, Novel ultrafiltration membranes from low-fouling copolymers for RO pretreatment applications, Desalination and Water Treat. 57 (2016) 23185-23. https://doi.org/10.1080/19443994.2016.1153906

[29] L. Zhang, P. Zhang, M. Wang, K. Yang, J. Liu, Research on the experiment of reservoir water treatment applying ultrafiltration membrane technology of different processes, J. Environ. Biol. 37 (2016)1007.

[30] D.L. Oatley-Radcliffe, M. Walters, T.J. Ainscough, P.M. Williams, A.W. Mohammad, N. Hilal, Nanofiltration membranes and processes: A review of research trends over the past decade, J. Water Process Eng. 19 (2017)164-171. https://doi.org/10.1016/j.jwpe.2017.07.026

[31] K.L., McMordie-Stoughton, X. Duan, E.M. Wendel, Reverse Osmosis Optimization, U.S. Department of Energy, 2013.

[32] R.H. Hailemariam, Y.C. Woo, M.M. Damtie, B.C. Kim, K.D. Park, J.S. Choi, Reverse osmosis membrane fabrication and modification technologies and future trends: A review, Adv. Colloid Interface Sci. 276 (2019) 102100. https://doi.org/10.1016/j.cis.2019.102100

[33] M. Asadollahi, D. Bastani, S.A. Musavi, Enhancement of surface properties and performance of reverse osmosis membranes after surface modification: a review. Desalination 420 (2017) 330 –383. https://doi.org/10.1016/j.desal.2017.05.027

[34] S. Ba, L. Haroune, L. Soumano, J.P. Bellenger, J.P. Jones, H. Cabana, A hybrid bioreactor based on insolubilized tyrosinase and laccase catalysis and microfiltration membrane remove pharmaceuticals from wastewater, Chemosphere 201 (2018) 749–755. https://doi.org/10.1016/j.chemosphere.2018.03.022

[35] L. Goswami, R.V. Kumar, K. Pakshirajan, G. Pugazhenthi, A novel integrated biodegradation—microfiltration system for sustainable wastewater treatment and energy recovery, J. Hazard. Mater. 365 (2019) 707–715. https://doi.org/10.1016/j.jhazmat.2018.11.029

[36] B. Jafari, M. Abbasi, S.A. Hashemifard, Development of new tubular ceramic microfiltration membranes by employing activated carbon in the structure of membranes for treatment of oily wastewater, J. Clean. Prod. 244 (2020) 118720. https://doi.org/10.1016/j.jclepro.2019.118720

[37] S. Saja, A. Bouazizi, B. Achiou, M. Ouammou, A. Albizane, J. Bennazha, S.A Younssi, Elaboration and characterization of low-cost ceramic membrane made from natural Moroccan perlite for treatment of industrial wastewater, J. Environ. Chem. Eng. 6 (2018) 451–458. https://doi.org/10.1016/j.jece.2017.12.004

[38] Z. Pan, F. Yu, L. Li, C. Song, J. Yang, C. Wang, Y. Pan, T. Wang, Electrochemical microfiltration treatment of bisphenol A wastewater using coal-based carbon

membrane, Sep. Purif. Technol. 227 (2019) 115695. https://doi.org/10.1016/j.seppur.2019.115695

[39] D. Beqqour, B. Achiou, A. Bouazizi, H. Ouaddari, H. Elomari, M. Ouammou, J. Bennazha, S. Alami Younssi, Enhancement of microfiltration performances of pozzolan membrane by incorporation of micronized phosphate and its application for industrial wastewater treatment, J. Environ. Chem. Eng. 7 (2019) 102981. https://doi.org/10.1016/j.jece.2019.102981

[40] M. Mouiya, A. Abourriche, A. Bouazizi, A. Benhammou, Y. El Hafiane, Y. Abouliatim, L. Nibou, M. Oumam, M. Ouammou, A. Smith, H. Hannache, Flat ceramic microfiltration membrane based on natural clay and Moroccan phosphate for desalination and industrial wastewater treatment, Desalination. 427 (2018) 42–50. https://doi.org/10.1016/j.desal.2017.11.005

[41] P. Arribas, M.C. García-Payo, M. Khayet, L. Gil, Heat-treated optimized polysulfone electrospun nanofibrous membranes for high performance wastewater microfiltration, Sep. Purif. Technol. 226 (2019) 323–336. https://doi.org/10.1016/j.seppur.2019.05.097

[42] C. Li, G. Feng, Z. Pan, C. Song, X. Fan, P. Tao, T. Wang, M. Shao, S. Zhao, High-performance electrocatalytic microfiltration CuO/Carbon membrane by facile dynamic electrodeposition for small-sized organic pollutants removal, J. Memb. Sci. 601 (2020) 117913. https://doi.org/10.1016/j.memsci.2020.117913

[43] M. Sheikhi, M. Arzani, H.R. Mahdavi, T. Mohammadi, Kaolinitic clay-based ceramic microfiltration membrane for oily wastewater treatment: Assessment of coagulant addition, Ceram. Int. 45 (2019) 17826–17836. https://doi.org/10.1016/j.ceramint.2019.05.354

[44] B. Hatimi, J. Mouldar, A. Loudiki, H. Hafdi, M. Joudi, E.M. Daoudi, H. Nasrellah, I.-T. Lançar, M.A. El Mhammedi, M. Bakasse, Low cost pyrrhotite ash/clay-based inorganic membrane for industrial wastewaters treatment, J. Environ. Chem. Eng. 8 (2020) 103646. https://doi.org/10.1016/j.jece.2019.103646

[45] A. Manni, B. Achiou, A. Karim, A. Harrati, C. Sadik, M. Ouammou, S. Alami Younssi, A. El Bouari, New low-cost ceramic microfiltration membrane made from natural magnesite for industrial wastewater treatment, J. Environ. Chem. Eng. 8 (2020) 103906. https://doi.org/10.1016/j.jece.2020.103906

[46] X. Zhang, B. Zhang, Y. Wu, T. Wang, J. Qiu, Preparation and characterization of a diatomite hybrid microfiltration carbon membrane for oily wastewater treatment, J. Taiwan Inst. Chem. Eng. 89 (2018) 39–48. https://doi.org/10.1016/j.jtice.2018.04.035

[47] N.C. Homem, N. de Camargo Lima Beluci, S. Amorim, R. Reis, A.M.S. Vieira, M.F. Vieira, R. Bergamasco, M.T.P. Amorim, Surface modification of a polyethersulfone microfiltration membrane with graphene oxide for reactive dyes removal, Appl. Surf. Sci. 486 (2019) 499–507. https://doi.org/10.1016/j.apsusc.2019.04.276

[48] M. Mouiya, A. Bouazizi, A. Abourriche, A. Benhammou, Y. El Hafiane, M. Ouammou, Y. Abouliatim, S.A. Younssi, A. Smith, H. Hannache, Fabrication and characterization of a ceramic membrane from clay and banana peel powder: Application to industrial wastewater treatment, Mater. Chem. Phys. 227 (2019) 291–301. https://doi.org/10.1016/j.matchemphys.2019.02.011

[49] X. Liu, B. Jiang, X. Yin, H. Ma, B.S. Hsiao, Highly permeable nanofibrous composite microfiltration membranes for removal of nanoparticles and heavy metal ions, Sep. Purif. Technol. 233 (2020) 115976. https://doi.org/10.1016/j.seppur.2019.115976

[50] H. Abadikhah, J.W. Wang, X. Xu, S. Agathopoulos, SiO_2 nanoparticles modified Si3N4 hollow fiber membrane for efficient oily wastewater microfiltration, J. Water Process Eng. 29 (2019) 100799. https://doi.org/10.1016/j.jwpe.2019.100799

[51] M. Chen, L. Zhu, J. Chen, F. Yang, C.Y. Tang, M.D. Guiver, Y. Dong, Spinel-based ceramic membranes coupling solid sludge recycling with oily wastewater treatment, Water Res. 169 (2020) 115180. https://doi.org/10.1016/j.watres.2019.115180

[52] K.V. Plakas, A. Mantza, S.D. Sklari, V.T. Zaspalis, A.J. Karabelas, Heterogeneous Fenton-like oxidation of pharmaceutical diclofenac by a catalytic iron-oxide ceramic microfiltration membrane, Chem. Eng. J. 373 (2019) 700–708. https://doi.org/10.1016/j.cej.2019.05.092

[53] P. Bhattacharya, D. Mukherjee, N. Deb, S. Swarnakar, S. Banerjee, Application of green synthesized ZnO nanoparticle coated ceramic ultrafiltration membrane for remediation of pharmaceutical components from synthetic water: Reusability assay of treated water on seed germination, J. Environ. Chem. Eng. 8 (2020) 103803. https://doi.org/10.1016/j.jece.2020.103803

[54] Z. Isik, E.B. Arikan, H.D. Bouras, N. Dizge, Bioactive ultrafiltration membrane manufactured from Aspergillus carbonarius M333 filamentous fungi for treatment of

real textile wastewater, Bioresour. Technol. Reports. 5 (2019) 212–219. https://doi.org/10.1016/j.biteb.2019.01.020

[55] P. Bhattacharya, D. Mukherjee, S. Dey, S. Ghosh, S. Banerjee, Development and performance evaluation of a novel CuO/TiO 2 ceramic ultrafiltration membrane for ciprofloxacin removal, Mater. Chem. Phys. 229 (2019) 106–116. https://doi.org/10.1016/j.matchemphys.2019.02.094

[56] G. Derouich, S. Alami Younssi, J. Bennazha, J.A. Cody, M. Ouammou, M. El Rhazi, Development of low-cost polypyrrole/sintered pozzolan ultrafiltration membrane and its highly efficient performance for congo red dye removal, J. Environ. Chem. Eng. 8 (2020) 103809. https://doi.org/10.1016/j.jece.2020.103809

[57] T. Li, W. Zhang, S. Zhai, G. Gao, J. Ding, W. Zhang, Y. Liu, X. Zhao, B. Pan, L. Lv, Efficient removal of nickel(II) from high salinity wastewater by a novel PAA/ZIF-8/PVDF hybrid ultrafiltration membrane, Water Res. 143 (2018) 87–98. https://doi.org/10.1016/j.watres.2018.06.031

[58] M. Shakak, R. Rezaee, A. Maleki, A. Jafari, M. Safari, B. Shahmoradi, H. Daraei, S.M. Lee, Synthesis and characterization of nanocomposite ultrafiltration membrane (PSF/PVP/SiO2) and performance evaluation for the removal of amoxicillin from aqueous solutions, Environ. Technol. Innov. 17 (2020) 100529. https://doi.org/10.1016/j.eti.2019.100529

[59] W. Ye, K. Ye, F. Lin, H. Liu, M. Jiang, J. Wang, R. Liu, J. Lin, Enhanced fractionation of dye/salt mixtures by tight ultrafiltration membranes via fast bio-inspired co-deposition for sustainable textile wastewater management, Chem. Eng. J. 379 (2020) 122321. https://doi.org/10.1016/j.cej.2019.122321

[60] H. Isawi, Evaluating the performance of different nano-enhanced ultrafiltration membranes for the removal of organic pollutants from wastewater, J. Water Process Eng. 31 (2019) 100833. https://doi.org/10.1016/j.jwpe.2019.100833

[61] C. Yang, W. Xu, Y. Nan, Y. Wang, Y. Hu, C. Gao, X. Chen, Fabrication and characterization of a high performance polyimide ultrafiltration membrane for dye removal, J. Colloid Interface Sci. 562 (2020) 589–597. https://doi.org/10.1016/j.jcis.2019.11.075

[62] L. Liu, Y. Xu, K. Wang, K. Li, L. Xu, J. Wang, J. Wang, Fabrication of a novel conductive ultrafiltration membrane and its application for electrochemical removal of hexavalent chromium, J. Memb. Sci. 584 (2019) 191–201. https://doi.org/10.1016/j.memsci.2019.05.018

[63] M.S. Algamdi, I.H. Alsohaimi, J. Lawler, H.M. Ali, A.M. Aldawsari, H.M.A. Hassan, Fabrication of graphene oxide incorporated polyethersulfone hybrid ultrafiltration membranes for humic acid removal, Sep. Purif. Technol. 223 (2019) 17–23. https://doi.org/10.1016/j.seppur.2019.04.057

[64] S. Saja, A. Bouazizi, B. Achiou, H. Ouaddari, A. Karim, M. Ouammou, A. Aaddane, J. Bennazha, S. Alami Younssi, Fabrication of low-cost ceramic ultrafiltration membrane made from bentonite clay and its application for soluble dyes removal, J. Eur. Ceram. Soc. 40 (2020) 2453–2462. https://doi.org/10.1016/j.jeurceramsoc.2020.01.057

[65] S. Arefi-Oskoui, A. Khataee, M. Safarpour, V. Vatanpour, Modification of polyethersulfone ultrafiltration membrane using ultrasonic-assisted functionalized MoS2 for treatment of oil refinery wastewater, Sep. Purif. Technol. 238 (2020) 116495. https://doi.org/10.1016/j.seppur.2019.116495

[66] G.P.S. Ibrahim, A.M. Isloor, Inamuddin, A.M. Asiri, A.F. Ismail, R. Kumar, M.I. Ahamed, Performance intensification of the polysulfone ultrafiltration membrane by blending with copolymer encompassing novel derivative of poly(styrene-co-maleic anhydride) for heavy metal removal from wastewater, Chem. Eng. J. 353 (2018) 425–435. https://doi.org/10.1016/j.cej.2018.07.098

[67] X. Huang, C. Tian, H. Qin, W. Guo, P. Gao, H. Xiao, Preparation and characterization of Al3+-doped TiO2 tight ultrafiltration membrane for efficient dye removal, Ceram. Int. 46 (2020) 4679–4689. https://doi.org/10.1016/j.ceramint.2019.10.199

[68] S. Benkhaya, B. Achiou, M. Ouammou, J. Bennazha, S. Alami Younssi, S. M'rabet, A. El Harfi, Preparation of low-cost composite membrane made of polysulfone/polyetherimide ultrafiltration layer and ceramic pozzolan support for dyes removal, Mater. Today Commun. 19 (2019) 212–219. https://doi.org/10.1016/j.mtcomm.2019.02.002

[69] S. Yu, X. Zhang, F. Li, X. Zhao, Poly(vinyl pyrrolidone) modified poly(vinylidene fluoride) ultrafiltration membrane via a two-step surface grafting for radioactive wastewater treatment, Sep. Purif. Technol. 194 (2018) 404–409. https://doi.org/10.1016/j.seppur.2017.10.051

[70] S. Benkhaya, S. M'rabet, R. Hsissou, A. El Harfi, Synthesis of new low-cost organic ultrafiltration membrane made from Polysulfone/Polyetherimide blends and its application for soluble azoic dyes removal, J. Mater. Res. Technol. 9 (2020) 4763-4772. https://doi.org/10.1016/j.jmrt.2020.02.102

[71] T. Ahmad, C. Guria, A. Mandal, Synthesis, characterization and performance studies of mixed-matrix poly(vinyl chloride)-bentonite ultrafiltration membrane for the treatment of saline oily wastewater, Process Saf. Environ. Prot. 116 (2018) 703–717. https://doi.org/10.1016/j.psep.2018.03.033

[72] M. Malmali, J. Askegaard, K. Sardari, S. Eswaranandam, A. Sengupta, S.R. Wickramasinghe, Evaluation of ultrafiltration membranes for treating poultry processing wastewater, J. Water Process Eng. 22 (2018) 218–226. https://doi.org/10.1016/j.jwpe.2018.02.010

[73] C. Cojocaru, L. Clima, Polymer assisted ultrafiltration of AO7 anionic dye from aqueous solutions: Experimental design, multivariate optimization, and molecular docking insights, J. Memb. Sci. 604 (2020) 118054. https://doi.org/10.1016/j.memsci.2020.118054

[74] S. Zhao, H. Zhu, Z. Wang, P. Song, M. Ban, X. Song, A loose hybrid nanofiltration membrane fabricated via chelating-assisted in-situ growth of Co/Ni LDHs for dye wastewater treatment, Chem. Eng. J. 353 (2018) 460–471. https://doi.org/10.1016/j.cej.2018.07.081

[75] S. Zhao, P. Song, Z. Wang, H. Zhu, The PEGylation of plant polyphenols/polypeptide-mediated loose nanofiltration membrane for textile wastewater treatment and antibacterial application, J. Taiwan Inst. Chem. Eng. 82 (2018) 42–55. https://doi.org/10.1016/j.jtice.2017.11.005

[76] Q. Zhang, S. Chen, X. Fan, H. Zhang, H. Yu, X. Quan, A multifunctional graphene-based nanofiltration membrane under photo-assistance for enhanced water treatment based on layer-by-layer sieving, Appl. Catal. B Environ. 224 (2018) 204–213. https://doi.org/10.1016/j.apcatb.2017.10.016

[77] B. Maryam, V. Buscio, S.U. Odabasi, H. Buyukgungor, A study on behavior, interaction and rejection of Paracetamol, Diclofenac and Ibuprofen (PhACs) from wastewater by nanofiltration membranes, Environ. Technol. Innov. 18 (2020) 100641. https://doi.org/10.1016/j.eti.2020.100641

[78] M.T. Tsehaye, J. Wang, J. Zhu, S. Velizarov, B. Van der Bruggen, Development and characterization of polyethersulfone-based nanofiltration membrane with stability to hydrogen peroxide, J. Memb. Sci. 550 (2018) 462–469. https://doi.org/10.1016/j.memsci.2018.01.022

[79] T.D. Kusworo, N. Ariyanti, D.P. Utomo, Effect of nano-TiO2 loading in polysulfone membranes on the removal of pollutant following natural-rubber wastewater

treatment, J. Water Process Eng. 35 (2020) 101190. https://doi.org/10.1016/j.jwpe.2020.101190

[80] D. Guo, Y. Xiao, T. Li, Q. Zhou, L. Shen, R. Li, Y. Xu, H. Lin, Fabrication of high-performance composite nanofiltration membranes for dye wastewater treatment: mussel-inspired layer-by-layer self-assembly, J. Colloid Interface Sci. 560 (2020) 273–283. https://doi.org/10.1016/j.jcis.2019.10.078

[81] S. Bandehall, F. Parvizian, A. Moghadassi, S.M. Hosseini, High water permeable PEI nanofiltration membrane modified by L-cysteine functionalized POSS nanoparticles with promoted antifouling/separation performance, Sep. Purif. Technol. 237 (2020) 116361. https://doi.org/10.1016/j.seppur.2019.116361

[82] C.Y. Wang, W.J. Zeng, T.T. Jiang, X. Chen, X.L. Zhang, Incorporating attapulgite nanorods into graphene oxide nanofiltration membranes for efficient dyes wastewater treatment, Sep. Purif. Technol. 214 (2019) 21–30. https://doi.org/10.1016/j.seppur.2018.04.079

[83] S.M. Abtahi, L. Marbelia, A.Y. Gebreyohannes, P. Ahmadiannamini, C. Joannis-Cassan, C. Albasi, W.M. de Vos, I.F.J. Vankelecom, Micropollutant rejection of annealed polyelectrolyte multilayer based nanofiltration membranes for treatment of conventionally-treated municipal wastewater, Sep. Purif. Technol. 209 (2019) 470–481. https://doi.org/10.1016/j.seppur.2018.07.071

[84] D.I. De Souza, E.M. Dottein, A. Giacobbo, M.A. Siqueira Rodrigues, M.N. De Pinho, A.M. Bernardes, Nanofiltration for the removal of norfloxacin from pharmaceutical effluent, J. Environ. Chem. Eng. 6 (2018) 6147–6153. https://doi.org/10.1016/j.jece.2018.09.034

[85] C. Wang, Y. Feng, J. Chen, X. Bai, L. Ren, C. Wang, K. Huang, H. Wu, Nanofiltration membrane based on graphene oxide crosslinked with zwitterion-functionalized polydopamine for improved performances, J. Taiwan Inst. Chem. Eng. 110 (2020) 153–162. https://doi.org/10.1016/j.jtice.2020.03.009

[86] K. Sangeetha, A.V. P., P.N. Sudha, A. Faleh A., A. Sukumaran, Novel chitosan based thin sheet nanofiltration membrane for rejection of heavy metal chromium, Int. J. Biol. Macromol. 132 (2019) 939–953. https://doi.org/10.1016/j.ijbiomac.2019.03.244

[87] H. Qin, W. Guo, X. Huang, P. Gao, H. Xiao, Preparation of yttria-stabilized ZrO2 nanofiltration membrane by reverse micelles-mediated sol-gel process and its

application in pesticide wastewater treatment, J. Eur. Ceram. Soc. 40 (2020) 145–154. https://doi.org/10.1016/j.jeurceramsoc.2019.09.023

[88] X.L. Cao, Y.N. Yan, F.Y. Zhou, S.P. Sun, Tailoring nanofiltration membranes for effective removing dye intermediates in complex dye-wastewater, J. Memb. Sci. 595 (2020) 117476. https://doi.org/10.1016/j.memsci.2019.117476

[89] T. Tavangar, M. Karimi, M. Rezakazemi, K.R. Reddy, T.M. Aminabhavi, Textile waste, dyes/inorganic salts separation of cerium oxide-loaded loose nanofiltration polyethersulfone membranes, Chem. Eng. J. 385 (2020) 123787. https://doi.org/10.1016/j.cej.2019.123787

[90] T. Yun, J.W. Chung, S.Y. Kwak, Recovery of sulfuric acid aqueous solution from copper-refining sulfuric acid wastewater using nanofiltration membrane process, J. Environ. Manage. 223 (2018) 652–657. https://doi.org/10.1016/j.jenvman.2018.05.069

[91] R. Pang, K. Zhang, Fabrication of hydrophobic fluorinated silica-polyamide thin film nanocomposite reverse osmosis membranes with dramatically improved salt rejection, J. Colloid Interface Sci. 510 (2018) 127–132. https://doi.org/10.1016/j.jcis.2017.09.062

[92] T. Fujioka, H. Aizawa, H. Kodamatani, Fouling substances causing variable rejection of a small and uncharged trace organic chemical by reverse osmosis membranes, Environ. Technol. Innov. 17 (2020) 100576. https://doi.org/10.1016/j.eti.2019.100576

[93] S.M. Ghaseminezhad, M. Barikani, M. Salehirad, Development of graphene oxide-cellulose acetate nanocomposite reverse osmosis membrane for seawater desalination, Compos. Part B Eng. 161 (2019) 320–327. https://doi.org/10.1016/j.compositesb.2018.10.079

[94] T. Fujioka, K.P. Ishida, T. Shintani, H. Kodamatani, High rejection reverse osmosis membrane for removal of N-nitrosamines and their precursors, Water Res. 131 (2018) 45–51. https://doi.org/10.1016/j.watres.2017.12.025

[95] E. Sahinkaya, S. Tuncman, I. Koc, A.R. Guner, S. Ciftci, A. Aygun, S. Sengul, Performance of a pilot-scale reverse osmosis process for water recovery from biologically-treated textile wastewater, J. Environ. Manage. 249 (2019) 109382. https://doi.org/10.1016/j.jenvman.2019.109382

[96] S. Qi, W. Fang, W. Siti, W. Widjajanti, X. Hu, R. Wang, Polymersomes-based high-performance reverse osmosis membrane for desalination, J. Memb. Sci. 555 (2018) 177–184. https://doi.org/10.1016/j.memsci.2018.03.052

[97] A. Egea-Corbacho Lopera, S. Gutiérrez Ruiz, J.M. Quiroga Alonso, Removal of emerging contaminants from wastewater using reverse osmosis for its subsequent reuse: Pilot plant, J. Water Process Eng. 29 (2019) 100800. https://doi.org/10.1016/j.jwpe.2019.100800

[98] P.D.A. Bastos, M.A. Santos, P.J. Carvalho, J.G. Crespo, Reverse osmosis performance on stripped phenolic sour water treatment – A study on the effect of oil and grease and osmotic pressure, J. Environ. Manage. 261 (2020) 110229. https://doi.org/10.1016/j.jenvman.2020.110229

[99] J.M. Ochando-Pulido, A. Martinez-Ferez, Optimization of the fouling behaviour of a reverse osmosis membrane for purification of olive-oil washing wastewater, Process Saf. Environ. Prot. 114 (2018) 323–333. https://doi.org/10.1016/j.psep.2018.01.004

[100] Q. Gu, T.C.A. Ng, I. Zain, X. Liu, L. Zhang, Z. Zhang, Z. Lyu, Z. He, H.Y. Ng, J. Wang, Chemical-grafting of graphene oxide quantum dots (GOQDs) onto ceramic microfiltration membranes for enhanced water permeability and anti-organic fouling potential, Appl. Surf. Sci. 502 (2020) 144128. https://doi.org/10.1016/j.apsusc.2019.144128

[101] L. Ghalamchi, S. Aber, V. Vatanpour, M. Kian, Comparison of NLDH and g-C3N4 nanoplates and formative Ag3PO4 nanoparticles in PES microfiltration membrane fouling: Applications in MBR, Chem. Eng. Res. Des. 147 (2019) 443–457. https://doi.org/10.1016/j.cherd.2019.05.033

[102] B. Kose-Mutlu, T. Turken, M.C. Guclu, S. Guclu, S. Ovez, I. Koyuncu, Effects of the post-modification using bismuth chelate (BisBAL) on the anti-biofouling and performance properties of flat-sheet microfiltration membranes, J. Water Process Eng. 23 (2018) 75–83. https://doi.org/10.1016/j.jwpe.2018.03.001

[103] J. Yang, F. Sun, L. Zhao, D.Y. Xing, W. Dong, Z. Dong, High-conductivity microfiltration membranes incorporated with ionic liquids and their superior anti-fouling effectiveness, J. Memb. Sci. 603 (2020) 117767. https://doi.org/10.1016/j.memsci.2019.117767

[104] Q. Gu, T.C.A. Ng, L. Zhang, Z. Lyu, Z. Zhang, H.Y. Ng, J. Wang, Interfacial diffusion assisted chemical deposition (ID-CD) for confined surface modification of

alumina microfiltration membranes toward high-flux and anti-fouling, Sep. Purif. Technol. 235 (2020) 116177. https://doi.org/10.1016/j.seppur.2019.116177

[105] X. Zhang, M. Ping, Z. Wu, C.Y. Tang, Z. Wang, Microfiltration membranes modified by silver-decorated biomimetic silica nanopollens for mitigating biofouling: Synergetic effects of nanopollens and silver nanoparticles, J. Memb. Sci. 597 (2020) 117773. https://doi.org/10.1016/j.memsci.2019.117773

[106] X. Zhang, Z. Wang, C.Y. Tang, J. Ma, M. Liu, M. Ping, M. Chen, Z. Wu, Modification of microfiltration membranes by alkoxysilane polycondensation induced quaternary ammonium compounds grafting for biofouling mitigation, J. Memb. Sci. 549 (2018) 165–172. https://doi.org/10.1016/j.memsci.2017.12.004

[107] X. Liu, H. Yuan, C. Wang, S. Zhang, L. Zhang, X. Liu, F. Liu, X. Zhu, S. Rohani, C. Ching, J. Lu, A novel PVDF/PFSA-g-GO ultrafiltration membrane with enhanced permeation and antifouling performances, Sep. Purif. Technol. 233 (2020) 116038. https://doi.org/10.1016/j.seppur.2019.116038

[108] Z. He, T.C.A. Ng, Z. Lyu, Q. Gu, L. Zhang, H.Y. Ng, J. Wang, Alumina double-layered ultrafiltration membranes with enhanced water flux, Colloids Surfaces A Physicochem. Eng. Asp. 587 (2020) 124324. https://doi.org/10.1016/j.colsurfa.2019.124324

[109] L. Zhang, T.C.A. Ng, X. Liu, Q. Gu, Y. Pang, Z. Zhang, Z. Lyu, Z. He, H.Y. Ng, J. Wang, Hydrogenated TiO2 membrane with photocatalytically enhanced anti-fouling for ultrafiltration of surface water, Appl. Catal. B Environ. 264 (2020) 118528. https://doi.org/10.1016/j.apcatb.2019.118528

[110] X. Huang, J. Zhang, K. Peng, Y. Na, Y. Xiong, W. Liu, J. Liu, L. Lu, S. Li, Functional magnetic nanoparticles for enhancing ultrafiltration of waste cutting emulsions by significantly increasing flux and reducing membrane fouling, 573 (2019) 73-84. https://doi.org/10.1016/j.memsci.2018.11.074

[111] F. Khoerunnisa, W. Rahmah, B. Seng Ooi, E. Dwihermiati, N. Nashrah, S. Fatimah, Y.G. Ko, E.-P. Ng, Chitosan/PEG/MWCNT/Iodine composite membrane with enhanced antibacterial properties for dye wastewater treatment, J. Environ. Chem. Eng. 8 (2020) 103686. https://doi.org/10.1016/j.jece.2020.103686

[112] N. Ahmad, A. Samavati, N.A.H.M. Nordin, J. Jaafar, A.F. Ismail, N.A.N.N. Malek, Enhanced performance and antibacterial properties of amine-functionalized ZIF-8-decorated GO for ultrafiltration membrane, Sep. Purif. Technol. 239 (2020) 116554. https://doi.org/10.1016/j.seppur.2020.116554

[113] U. Sathya, M. Nithya, Keerthi, Fabrication and characterisation of fine-tuned Polyetherimide (PEI)/WO3 composite ultrafiltration membranes for antifouling studies, Chem. Phys. Lett. 744 (2020) 137201. https://doi.org/10.1016/j.cplett.2020.137201

[114] G. Zhang, M. Zhou, Z. Xu, C. Jiang, C. Shen, Q. Meng, Guanidyl-functionalized graphene/polysulfone mixed matrix ultrafiltration membrane with superior permselective, antifouling and antibacterial properties for water treatment, J. Colloid Interface Sci. 540 (2019) 295–305. https://doi.org/10.1016/j.jcis.2019.01.050

[115] R.P. Pandey, P.A. Rasheed, T. Gomez, R.S. Azam, K.A. Mahmoud, A fouling-resistant mixed-matrix nanofiltration membrane based on covalently cross-linked $Ti_3C_2T_X$ (MXene)/cellulose acetate, J. Memb. Sci. 607 (2020) 118139. https://doi.org/10.1016/j.memsci.2020.118139

[116] Y.F. Mi, G. Xu, Y.S. Guo, B. Wu, Q.F. An, Development of antifouling nanofiltration membrane with zwitterionic functionalized monomer for efficient dye/salt selective separation, J. Memb. Sci. 601 (2020) 117795. https://doi.org/10.1016/j.memsci.2019.117795

[117] H. Koulivand, A. Shahbazi, V. Vatanpour, M. Rahmandoust, Development of carbon dot-modified polyethersulfone membranes for enhancement of nanofiltration, permeation and antifouling performance, Sep. Purif. Technol. 230 (2020) 115895. https://doi.org/10.1016/j.seppur.2019.115895

[118] W. Shang, F. Sun, W. Jia, J. Guo, S. Yin, P.W. Wong, A.K. An, High-performance nanofiltration membrane structured with enhanced stripe nano-morphology, J. Memb. Sci. 600 (2020) 117852. https://doi.org/10.1016/j.memsci.2020.117852

[119] V. Vatanpour, N. Haghighat, Improvement of polyvinyl chloride nanofiltration membranes by incorporation of multiwalled carbon nanotubes modified with triethylenetetramine to use in treatment of dye wastewater, J. Environ. Manage. 242 (2019) 90–97. https://doi.org/10.1016/j.jenvman.2019.04.060

[120] X. Zhu, X. Tang, X. Luo, X. Cheng, D. Xu, Z. Gan, W. Wang, L. Bai, G. Li, H. Liang, Toward enhancing the separation and antifouling performance of thin-film composite nanofiltration membranes: A novel carbonate-based preoccupation strategy, J. Colloid Interface Sci. 571 (2020) 155-165. https://doi.org/10.1016/j.jcis.2020.03.044

[121] S. Li, B. Gao, Y. Wang, B. Jin, Q. Yue, Z. Wang, Antibacterial thin film nanocomposite reverse osmosis membrane by doping silver phosphate loaded

graphene oxide quantum dots in polyamide layer, Desalination. 464 (2019) 94–104. https://doi.org/10.1016/j.desal.2019.04.029

[122] Z. Yang, D. Saeki, R. Takagi, H. Matsuyama, Improved anti-biofouling performance of polyamide reverse osmosis membranes modified with a polyampholyte with effective carboxyl anion and quaternary ammonium cation ratio, J. Memb. Sci. 595 (2020) 117529. https://doi.org/10.1016/j.memsci.2019.117529

[123] H. Lee, G. Amy, J. Cho, Y. Yoon, S.-H. Moon, I.S. Kim, Cleaning strategies for flx recovery of an ultrafitration membrane fouled by natural organic matter, Water Res. 35 (2001) 3301–3308. https://doi.org/10.1016/S0043-1354(01)00063-X

[124] R. Liikanen, J. Yli-Kuivila, R. Laukkanen, Effiency of various chemical cleanings for nanofitration membrane fouled by conventionally-treated surface water, J. Membr. Sci. 195 (2002) 265–276. https://doi.org/10.1016/S0376-7388(01)00569-5

[125] M. Rabiller-Baudry, L. Bégoin, D. Delaunay, L. Paugam, B. Chaufer, A dual approach of membrane cleaning based on physico-chemistry and hydrodynamics: application to PES membrane of dairy industry, Chem. Eng. Process. Process Intensif. 47 (2008) 267–275. https://doi.org/10.1016/j.cep.2007.01.026

[126] S.A. Aktij, A. Taghipour, A. Rahimpour, A. Mollahosseini, A. Tiraferri, A Critical Review on Ultrasonic-Assisted Fouling Control and Cleaning of Fouled Membranes. Ultrasonics, 108 (2020) 106228. https://doi.org/10.1016/j.ultras.2020.106228

[127] E. Alventosa-deLara, S. Barredo-Damas, M. Alcaina-Miranda, M. Iborra-Clar, Study and optimization of the ultrasound-enhanced cleaning of an ultrafitration ceramic membrane through a combined experimental–statistical approach, Ultrason. Sonochem. 21 (2014) 1222–1234. https://doi.org/10.1016/j.ultsonch.2013.10.022

[128] A. Al-Amoudi, R.W. Lovitt, Fouling strategies and the cleaning system of NF membranes and factors affcting cleaning effiency, J. Membr. Sci. 303 (2007) 4–28. https://doi.org/10.1016/j.memsci.2007.06.002

[129] C. Shorrock, M. Bird, Membrane cleaning: chemically enhanced removal of deposits formed during yeast cell harvesting, Food Bioprod. Process. 76 (1998) 30–38. https://doi.org/10.1205/096030898531729

[130] N. Porcelli, S. Judd, Chemical cleaning of potable water membranes: a review, Sep. Purif. Technol. 71 (2010) 137–143. https://doi.org/10.1016/j.seppur.2009.12.007

[131] C. Liu, S. Caothien, J. Hayes, T. Caothuy, T. Otoyo, T. Ogawa, Membrane chemical cleaning: from art to science, Pall Corporation 11050 Port Washington, NY, 2001.

[132] E. Zondervan, B. Roffl, Evaluation of diffrent cleaning agents used for cleaning ultra fitration membranes fouled by surface water, J. Membr. Sci. 304 (2007) 40–49. https://doi.org/10.1016/j.memsci.2007.06.041

Materials Research Foundations **102** (2021) 106-127 | https://doi.org/10.21741/9781644901397-4

Chapter 4

Sustainable Organo-Inorganic Metal Composites for Catalytic Degradation of Industrial Persistent Toxic Substances and Energy Storage

Suranjana V. Mayani[1,2], Sang Wook Kim[1] and Vishal J. Mayani[1,3]*

[1]Department of Advanced Materials Chemistry, College of Science and Engineering, Dongguk University, Gyeongju Campus, Gyeongju 780714, Gyeongbuk, Republic of Korea

[2]Mayani Advanced-Chemicals Research Group (MARG), Department of Chemistry, Faculty of Science, Marwadi University, Rajkot 360003, Gujarat, India

[3]Hansgold ChemDiscovery Center (HCC), Hansgold ChemDiscoveries LLP, Rajkot 360005, Gujarat, India.

*vjmayani@gmail.com

Abstract

Colossal quantity of wastewater contaminated with persistent hazardous substances and degradable toxic compounds to the atmosphere are generated annually. Amongst the particular pollutant, chemicals, organic dyestuffs are of a considerable significance by virtue of its applications in fibers, fabric, coloring element, printed matter and manuring production. The present chapter comprises a complete perspective of green and sustainable organic-inorganic metal nanocomposites for heterogeneous chemical curtail of precarious organic noxious tinge (Chromotrope-2R, Eosin-Y and Methylene Blue) and energy storage (H_2 and C_2H_4). The nanocomposites were designed by simple strategy adopting nanostructured porous carbon material developed from reasonable pyrolysis fuel oil (PFO) related pitch remains.

Keywords

Economical, Energy Storage, Green Heterogeneous Catalyst, Hazardous Water Pollutants, Organic Dyes

Contents

1. Introduction

Water is a precious reservoir and not a single living creature can live without H_2O. The large quantity discharge of untreated sewage and industrial wastes are polluting the habitation and environment. Contemporary applied industries are practicing the use of organic colouring agents, pigments and dye materials globally. Polymer industry, foodstuff, garments, synthetic leather and pulp industries are the major stock holders of these significant dye materials. The pollutant clearance and toxic waste of these factories is contaminating the water and soil. Ubiquitously, environmental researchers are working on the counteraction of this colouring agent-associated waste materials and its reprocessing [1]. Methylthioninium chloride, also known as methylene blue (MB), is used in medication to treat methemoglobinemia but it is mainly promoted as a dying agent to colour cottons, wood and silk threads. The contamination of methylene blue dye is very harmful to the environmental health [2-4]. Eosin-Y is a xanthene group's red fluorescent pigment. Due to its stability, adaptability, and light absorption nature, it is broadly availed in diverse applications such as., printing, polymer production, sensitizing industry, printing ink, fluorescent dye compounds, nanomaterials, histology, and biomedical research [5-8]. The discharge of Eosin-Y effluents spoils the eco-environment system due to its deep colour and noxious properties. Its structural features, combination of carboxyl group, pyran and benzene, enables carcinogenic vulnerability as a dye waste. Thirdly, the union of aromatic ring, chromophore mono-azo (−N=N−), hydroxyl (−OH), sulfonyl (−SO_3H) functional groups composes unique dye which is represented as a Chromotrope 2R. Due to its

Materials Research Forum LLC
https://doi.org/10.21741/9781644901397-4

carcinogenic, destructive and venomous behaviour, the effluent of Chromotrope 2R causes serious environmental problems [1, 9-11]. Organic dyes in drain water experienced chemical changes, absorbs dissolved oxygen and ruins marine life. The stain and immense chemical oxygen demand (COD) of these discharges may cause serious natural problems. For future generation environment protection, it is necessary to decompose these pollutant materials. A competent lessening and curing capabilities are therefore required to discard the poisonous or cancerous pigment masses, these colouring agents are a real matter in the area of contaminated water therapy. A large number of scientific researchers have shaped firm attempts to develop various methods and techniques for elimination of dyes and persistent pollutants in drain water. Presently, diverse scientific methods are viable such as molecular concentration, solid eradication, micellar improved active penetration, micro-purification and chemical sorption on the surface of adsorbent, activated farming hard wastage, hybrid carbon matters, magnetic compounds and catalytic decomposition. Lately, several processes have been employed to eliminate or breakdown of the organic dyes from wastewater, e.g., agglomeration, sorption, catalytic redox reaction, living science processes and photocatalysis. These methods failed to produce total elimination or breakdown of organic dye pollutants; however, it offered certain hitches like partial removal, great consumption of chemicals and reagents, pricey, and creation of inferior toxic pollutants. [12]. These organic dye pollutants are chemically fragmentized to smaller molecular weight, simple and readily switchable materials such as oxygenated molecules, acetic acid, propionic acid, ethanol and minerals (CO_2, H_2O). These decomposed materials are non-toxic and can be easily released in the environment [13]. Nanostructured porous-carbon materials have exhibited huge potential for the elimination of pollutants by using progressive heterogeneous oxidation/reduction reactions. Sustainable, non-pricey and green metallic-carbon nanomaterial acquires united characteristics of basic-organic structure, metals and carbon-establishment together with the probability of customizing the activeness and preferences of nanomaterials. Addition of selective non-carbon/non-hydrogen atoms and transition metal into the nano-porous carbon framework will arouse the composite active sites. In the last decade, several reports of transition metal-associated composites have been published because of its immense fertile function as a solid-catalyst, chemical sorbent, carbon-electrode, power stockpile shell [14]. This chapter comprises a wide-ranging review of green and solid organic-inorganic metal nanocomposites for chemical breakdown of toxic organic dyes such as., Chromotrope-2R, Methylene Blue and Eosin-Y (Fig. 1) along with nanocomposite material's hydrogen and ethylene gas adsorption activity for energy storage. The nanocomposites were prepared by an easy technique using nano-sized porous carbon structure synthesized from cheaper petroleum pyrolysis fuel oil (PFO) based pitch excess. These sustainable catalysts have gigantic

Materials Research Forum LLC
https://doi.org/10.21741/9781644901397-4

prospective for the operation of purifying hetero-atom dye contaminated H_2O and it was freshly employed for novel sorption purpose as well.

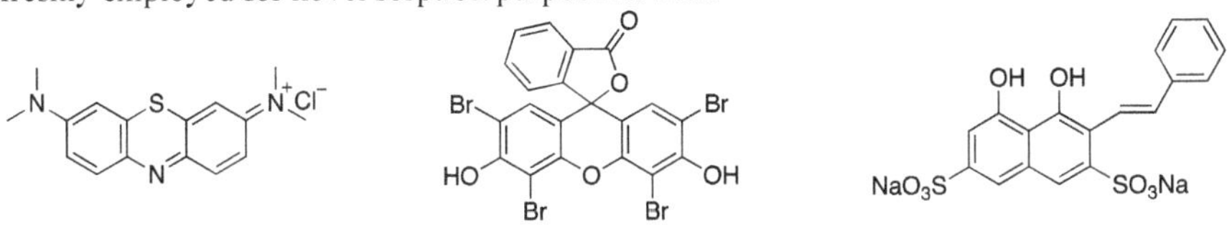

Methylene Blue   Eosin-Y   Chromotrope-2R

Figure 1. Chemical structures of Chromotrope-2R, Methylene Blue and Eosin-Y dyes.

2. Petroleum remnant as a carbon precursor

The world is greatly understanding the benefits of petroleum resources and its limited reserves. Low-cost petroleum pyrolysis fuel oil (PFO) is a higher boiling point multi-ring non-aliphatic hydrocarbon compound of crude oil residue that are supplied from petroleum distillation process or naphtha cracking process. Naphthalene is one of the binary-ring aromatic products and its derivatives are an essential part of various biologically important naturally occurring valuable chemicals, organic dyes and medicines [15-17]. Valuable naphthalene crystals were separated from PFO by developing innovative and economical solvent extraction methods. PFO based pitch was considered for the higher carbon content source after subsequent separation of valuable products [15, 18]. PFO pitches are applied extensively as precursors for porous carbon framework due to their finest graphitizable character, unique structural-shape, huge carbon amount, thermal flexibility and no burning remains. The particular qualities of carbon precursor are a promising applicant for the preparation of highly functioning carbon nanocomposite by applying a molding technique [19–21]. Schematic diagram of naphthalene separation and isolation of pitch as carbon precursor from PFO is shown in Fig. 2.

The nano silica ball (NSB), homogeneously uniform nano-sized silica sphere template, was prepared by the water based chemical breakdown and condensation of tetraethoxy orthosilicate (TEOS) [22]. Fresh solvent isolated PFO pitch excess was generated as per the report [15]. The prepared pitch paste (1.0 g) was taken in a rectangle-shaped ceramic crucible and heated at 140°C for half an hour until the formation of soft and highly moldable amber coloured semi-liquid. Uniform sized NSB (1.0 g) bed was mixed homogeneously with pitch fluid along with heating until the formation of dark amber coloured dust. The carbonization of the composite was carried out at 900°C in presence of nitrogen gas atmosphere in a tubular furnace to get nano silica carbon composite (CSC) for 4 hours. The nano-silica content from CSC was removed by stirring the mixture with aqueous 10% HF solution for 24 hours. Finally, the process was succeeded by filter-separation, H_2O washing and de-moisturization at 110°C for 4 hours to achieve a black colour nano carbon cage (NCC) with great carbon turnout (0.62 g). This nano carbon cage (NCC) can be used as a porous host for direct metal doping [18]. The covalent binding of ligand or metal complex makes hybrid composite sustainable and leaching independent. Therefore, the synthesized crude oil pitch originated carbon nano-cage (NCC) was further oxidized in combined presence of N_2 and 5% O_2 gas at 698 K for deep 10 hours to secure NCC-OH. The facet -OH moiety of NCC-OH was refluxed with watery sodium hydroxide (0.1N NaOH) solution for 1 hour under maintained pH condition to form NCC-ONa (Fig. 3). All the three nanocomposites NCC, NCC-OH, NCC-ONa can be directly used as a porous host for direct metal doping or covalent bonding [1].

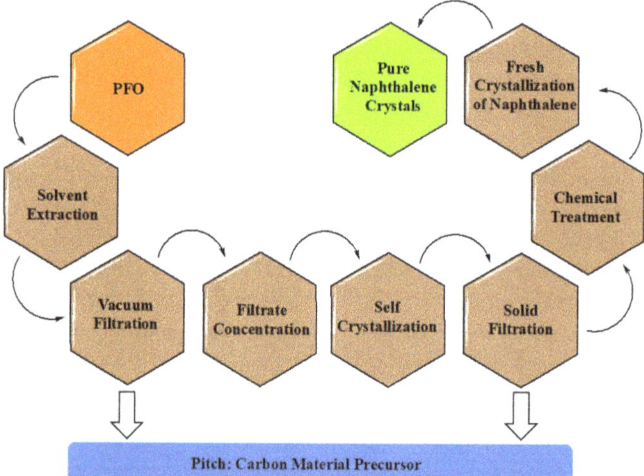

Figure 2. Schematic diagram of separation of carbon material precursor.

2.1. Organic-inorganic metal nanocomposites

Commercially, activated carbons are generally jumbled with irregular pore-sizes and they regularly have the impurities of the origin source, but nano-porous carbon materials prepared by template methods are highly ordered and have uniform pore size. The requirement of homogeneous metal carbon nanocomposites has been increased considerably in the last few years due to their broad applications in large scale industrial adsorption, gas isolation, water cleansing and heterogeneous catalysis. In continuation of our development of organic-inorganic hybrid catalyst and chiral stationary phases [23-25], the current study will deliver a comprehensive evaluation of synthesized carbon nanocomposites of gold (**Au@NCC**), copper (**Cu@NCC**), nickel (**Ni@NCC**), potassium and manganese (**K-Mn@NCC**), phosphorous (**P@NCC**), gold-phosphorous (**Au-P@NCC**), molybdenum-vanadium (**Mo-V@NCC**), palladium (**Pd@NCC**), gold-palladium (**Au-Pd@NCC**), tungsten (**W@NCC**) and covalently bonded gold Salen nanocomposite (**Au-Salen@NCC**) and its catalytic activities along with energy storage capacity. The schematic diagram of synthetic route of all the 11 organic-inorganic metal carbon nanocomposites is given in Fig. 4. The characterization of all nanocomposites and products have been accomplished by microanalysis, ^{1}H & ^{13}C NMR, ^{13}C CP-MAS NMR cross polarized magic angle spinning NMR, FTIR, GC, LCMS, UV-Visible spectroscopy, fine-powdered X-ray diffraction (XRD), thermal gravimetric analysis (TGA), surface analysis, scanning electron microscopy coupled energy dispersive X-ray spectroscopy (SEM/EDS), high resolution transmission electron microscopy (TEM), inductive coupled plasma (ICP) based metal analysis, and solid reflectance UV–vis spectroscopy.

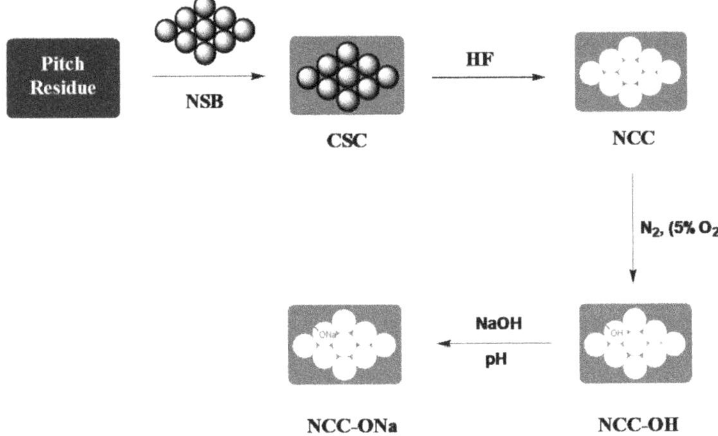

*Figure 3. Synthesis route of Nano carbon cage (**NCC**), Hydroxy nano carbon cage (**NCC-OH**) and Sodium hydroxy nano carbon cage (**NCC-ONa**).*

All eleven synthesized metal carbon nanocomposites, **Au@NCC** [26, 27], **Cu@NCC**, **Ni@NCC**, **K-Mn@NCC** [18, 27], **P@NCC**, **Au-P@NCC** [28], **Mo-V@NCC** [3], **Pd@NCC**, **Au-Pd@NCC** [29, 30], **W@NCC** [14], and **Au-Salen@NCC** [1], displays not only the singular physico-chemical essence of involved metal and nanoporous carbon cage (NCC) but also the collective characteristics of the paired composite and the occurrence of new qualities able of promoting brand-new operations (Fig. 4).

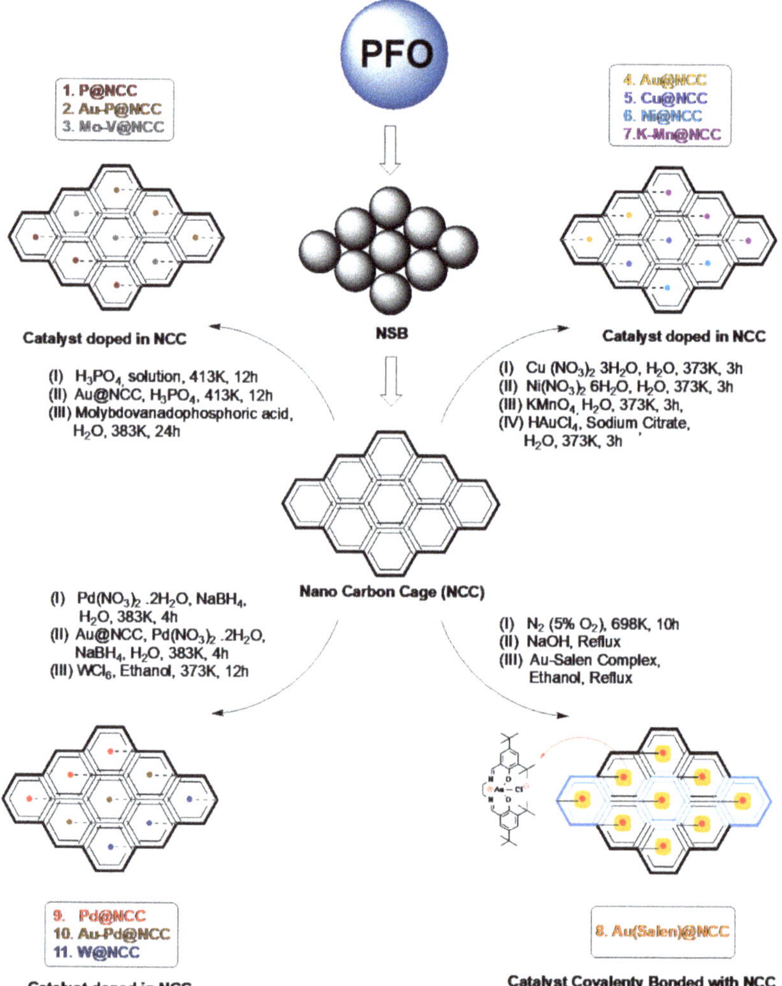

Figure 4. Synthesis route of eleven organic-inorganic metal nanocomposites ***1-11****.*

Materials Research Foundations **102** (2021) 106-127 https://doi.org/10.21741/9781644901397-4

3. Removal process of organic pollutant dyes

To expand the improved awareness of this materials, the following nanocomposites were successfully used as heterogeneous catalytic reaction of Eosin-Y, Methylene Blue, and Chromotrope-2R dyes. They were also utilized as an energy storage adsorbent for the adsorption of hydrogen gas and ethylene gas, and as an electrode for methanol oxidation reaction.

3.1 Catalytical degradation of methylene blue

The heterogeneous catalytic degradation of potentially destructive dye, methylene blue (MB), was effectively investigated by molybdenum-vanadium carbon nanocomposite (**Mo-V@NCC**). The catalytic degradation of MB dye was accomplished by taking 1 g of the recyclable nanocomposite catalyst **Mo-V@NCC** in round-based flask applying 50 ml of methylene blue solution (10^{-4} M, MB) and 50ml 30% hydrogen peroxide (H_2O_2, 10^{-4} M). The catalytic decomposition was observed by UV-visible and FT-IR spectroscopy [3].

The full range (200-800 nm) and selective (450-800 nm) UV-vis spectra of MB decomposition are revealed in Fig. 5 and 6. The UV-Vis spectrum of MB displayed major absorbance peak at 664 nm initially. The spectrum also shown three additional key projections at 250, 297 and 665 nm because of π to π^* electron alteration of the aromatic cycle and hetero atom multi-aromatic alliance in MB. The intensity of all the peaks was getting weaker and totally disappeared after 60 min catalytic reaction run with **Mo-V@NCC** catalyst. It denotes that the hazardous pigments were absolutely fragmented into small chemical segments. An FT-IR study of MB dye decomposition is show in Fig. 7. Originally, non-catalysed MB shown characteristic peaks of C=N stretching vibration, C=C vibrations, multiple ring stretching vibrations, C_{Ar}-N (the bond between the side aromatic ring and nitrogen atom) stretching vibration and N-CH_3 stretching vibrations 1601, 1544 & 1489, 1398, 1354, 1241 & 1181 cm^{-1} respectively [31]. Subsequently, original untreated MB characteristic bands were disappeared after catalytic degradation reaction which indicated the complete decomposition of pollutant dye. New additional peaks were generated at 1404, 1398 & 1225 cm^{-1}, these peaks mayhap as a result of IR-stretching vibrations of COO-, NO_3^- and SO_4^{2-} functional groups, respectively. Additionally, new peaks confirmed the presence of carbon dioxide as a chief final product along with some acids. FT-IR spectra also supported the complete oxidative decomposition of MB like UV-vis spectroscopy. The recovered **Mo-V@NCC** was recycled and reused for several catalytical run successfully with no noticeable loss in its activity [3].

Materials Research Forum LLC
https://doi.org/10.21741/9781644901397-4

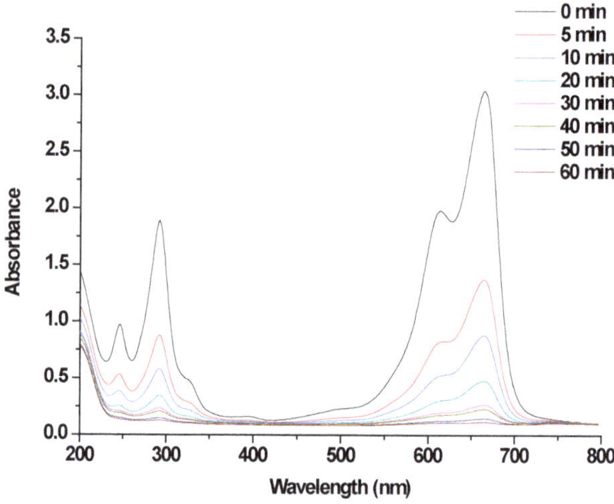

*Figure 5. Full range UV-vis spectrum of MB degradation by **Mo-V@NCC**.*

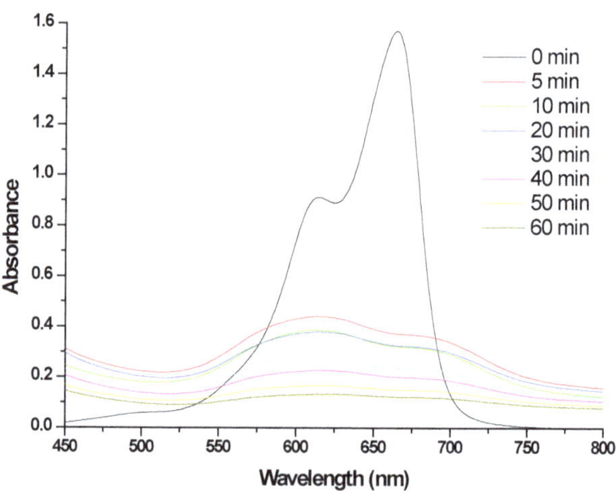

*Figure 6. Selective absorption UV-vis spectrum of MB degradation by **Mo-V@NCC**.*

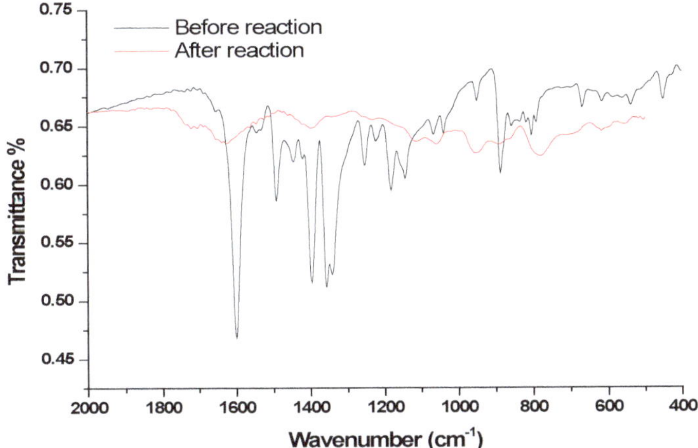

*Figure 7. FT-IR spectrum of MB afore and later degradation by catalysts **Mo-V@NCC**.*

Moreover, the rapid adsorption ability of **W@NCC** nanocomposite to methylene blue was detected by single outing the discoloration of MB dye content. This swift adsorption characteristics of **W@NCC** lead to the new catalytic oxidation study of MB. The full range UV-Visible spectra MB degradation using **W@NCC** is shown in Fig. 8. The spectrum shown dominant absorbance projections at 665 nm and binary little absorbance peaks at 251 and 297 nm. As we have seen in **Mo-V@NCC** case, similar phenomenon of the peak intensity weakening and disappearing was observed after 25 min vigorous reaction with **W@NCC**. It suggested that the aromatic and hetero-atom multi-aromatic link of MB dye was fully destroyed and entirely fragmented into scaled-down products. The catalytic oxidation of MB using **W@NCC** was also confirmed and supported by FT-IR spectroscopy [14].

Materials Research Foundations **102** (2021) 106-127
https://doi.org/10.21741/9781644901397-4

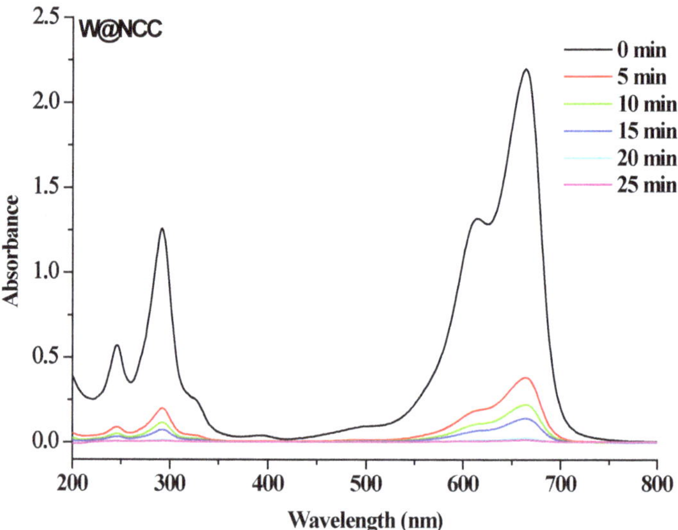

Figure 8. Full range UV-vis spectrum of MB degeneration by ***W@NCC****.*

3.2 Catalytical breakdown of chromotrope-2R and Eosin-Y

Jacobsen Salen Schiff base metal complexes have been largely studied because of their important catalytic activity and various organic reaction [32]. Together, several developed redox reactions of toxic waste healing have been rejuvenated amidst inexpensive, eco-friendly and huge productivity to settle the strong requirement of straight sewerage analysis. Less articles have been reported for the chemical reactionary removal Chromotrope-2R and Eosin-Y and from wastewater [1]. Au-Salen compound bonded nano-structured carbon material, **Au-Salen@NCC**, has assured the breakdown of Chromotrope-2R and Eosin-Y dyes with mild reaction modes employing modest oxidant hydrogen peroxide (H_2O_2) for Chromotrope-2R and reducing agent sodium borohydride ($NaBH_4$) for Eosin-Y and it is found to be highly significance and capable in contrast to available literature report [33].

The catalytic destruction and fragmentation of Eosin-Y lead the formation of simpler acid compounds which were further decomposed to essential mineral compounds. Gas chromatography study recognized four different products viz., (1Z,4E)-1,5-dibromo-3((Z)-

2carboxyvinyl)-6-oxohexa-1,4-dien-2-olate, 2-((2-3,6-dihydro-2H-pyran-4-yl) phenyl)2-carboxylate, 2-(2-formylphenyl)-2-carboxylate and 3,5 dibromocyclohex-5-ene-1,2,4-trione. The GC data was in support of Eosin-Y degradation mechanism [1]. These four intermediate compounds experienced subsequent decomposition to yield oxalic acid and malonic acid which were additionally degenerated to carbon dioxide and water. **Au-Salen@NCC** achieved 98.68 % degradation of Eosin Y within less than 1-hour time at room temperature and ordinary pressure, whilst other reported titanium dioxide (TiO_2) based sunlight-catalyst delivered 96 % decomposition yield under visible light treatment for 3 hours [33]. The Fig. 9 reveals the UV-vis spectrum of Eosin-Y decomposition using nanostructured **Au-Salen@NCC**. The UV-vis spectra of Eosin-Y identified main spectral-absorbance peak at 515 nm. The spectrum strongly suggests that the magnitude of spectral-absorbance bands evolved weaker as the reduction activity proceeded and peaks eliminated completely in 50 minutes.

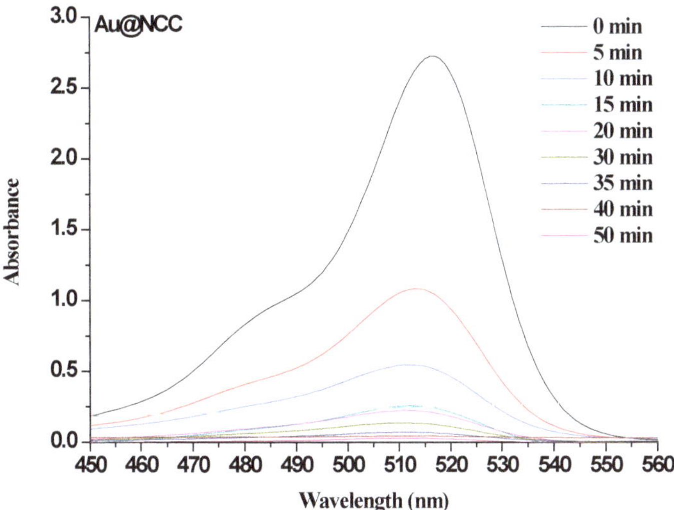

*Figure 9. Full range UV-vis spectrum of Eosin Y degeneration using **Au-Salen@NCC**.*

Catalytic degradation activity of Chromotrope-2R produced few capable acidic compounds as a result of decomposition of residual functional groups such as aromatic ring, −N = N−, −OH and $-SO_3H$ [34,35]. GC-MS analysis gave confirmative peaks for oxamic acid, oxalacetic acid, oxalic and malonic acid as the ultimate outputs that were break-downed to carbon dioxide [1]. Proportionately, 98.8% destruction of Chromotrope-2R was done by **Au-Salen@NCC** nanocomposite in period of one hour with respect to the reported Fenton process [34]. The present catalyst **Au-Salen@NCC** and technique was found to be very efficient, and eco-friendly and time saviour for full destruction of pollutant dyes. The UV-Visible investigation spectrum of Chromotrope 2R using **Au-Salen@NCC** nanocomposite is displayed in Fig. 10. The UV-vis spectrum of Chromotrope-2R dye displayed main absorbance at 508 nm. Usually, the disappearing of major absorbance band peak with course of oxidative catalytic reaction and complete removal after 55 min.

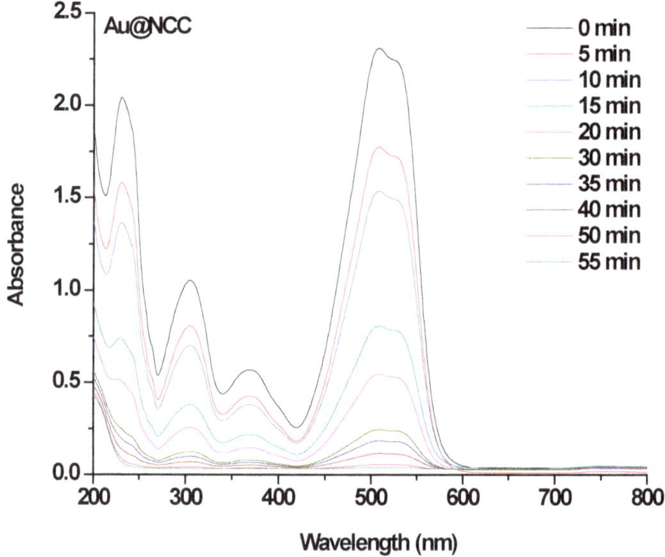

Figure 10. Full range UV-vis spectrum of Chromotrope-2R degeneration by ***Au-Salen@NCC****.*

The solid-catalyst optimization study of **Au-Salen@NCC** nanocomposite was conducted for both Chromotrope 2R and Eosin Y dyes (Table 1). The total conversion of dye treatment was increased reasonably during the catalyst load increment from 1, 3, 5, 8, 10 g in both the dyes. On the basis of research findings, the optimum catalyst amount (5 gL−1) and maximum conversion time (50 and 55) was found for Eosin-Y and Chromotrope-2R pigments, respectively. Recycled **Au-Salen@NCC** catalyst worked well for four repeated experiment without any leaching and performance loss. **Au-Salen@NCC** was sustainable and adequate compared to past literature reports.

Table 1. Solid-Catalyst load study on degradation of Chromotrope-2R and Eosin-Y.

Solid Catalyst loading (g L^{-1})	Eosin-Y	Chromotrope-2R
	Catalytic Reaction Time (min)	Catalytic Reaction Time (min)
1	90	80
3	70	70
5	50	55
8	50	55
10	50	55

4. Adsorption and energy storage

In the instant energy consumption era, the adsorption and storage of primary gases is one of the crucial concerns for the awareness for power maintenance and undoing the earth surface warming process viz., greenhouse. Few unhealthy volatile matters are supposed to be main suppliers to acidic rainfall, hazardous air pollution and elevated temperature of world. The diminishing fossil fuel reservoir and growing substantial deterioration knowledge have started an exploration for substitute fuel-materials and zero discharge carriers.

4.1 Hydrogen (H_2) gas adsorption study

Hydrogen gas is predicted to play a major part in the forthcoming world energy equilibrium. The efficient storage of H_2 gas is a challenging condition and need for its use as a vehicle fuel. The H_2 gas running cars have good advantage over battery and petrol cars. H_2 gas can be easily produced by electrolysis of H_2O, only release from its succeeding reaction with O_2 in the energy producing phase is H_2O. and H_2 gas has the maximum energy volume (120 MJ/kg) per mass unit [27, 36]. The H_2 gas adsorption isotherms were accomplished by using gravimetric adsorption system (MSB Rubotherm) in four particular stages: black analysis, stowing and activation of specimens, buoyancy study and sorption analysis. Later, the risen mass of the specimen by virtue of H_2 gas sorption was precisely counted by utilizing MSB connected to the sampling site. Normally, 0.2~0.7 g of adsorbent was taken for adsorption experiment. Adsorption isotherms metal carbon nanocomposites and its precursors, **Cu@NCC**, **Ni@NCC**, **K-Mn@NCC**, **Au@NCC** and non-metal pristine **NCC**, were studied to check the performance of metal-associated carbon nanocomposites. The hydrogen (H_2) gas sorption volume of nanocomposites is articulated in phrase of wt. % (mass of the hydrogen gas energy carrier/mass of the nanocomposite x 100) [27, 29].

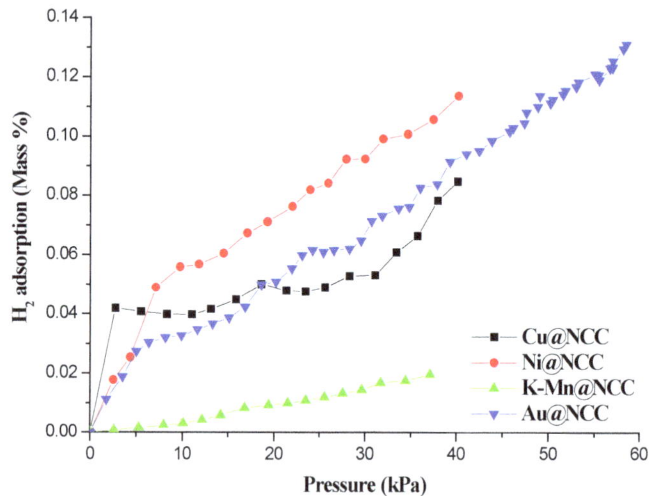

*Figure 11. Hydrogen adsorption of metal carbon nanocomposites **Cu@NCC**, **Ni@NCC**, **K-Mn@NCC** and **Au@NCC**.*

The hydrogen gas sorption study of pure NCC and its metal nanocomposites **Cu@NCC**, **Ni@NCC**, **K-Mn@NCC**, **Au@NCC** were investigated using standard gravimetric adsorption analysis. Nanocomposites **Cu@NCC** and **Ni@NCC** revealed highest H_2 gas adsorption capacity of 0.08 and 0.11 wt % around 40 kilopascal (kPa) pressure, respectively. Also, **Au@NCC** shown higher adsorption with raised pressure (0.13 wt %, 58.5 kPa). While **K-Mn@NCC** isotherm exhibited minimal H_2 gas with pressure rise (0.02 wt %, 37.1 kPa) compare to other nanocomposites (Fig. 11). Contrarily, active H_2 gas sorption was missing for pure NCC sample at physical force up to 5000 kilopascal (kPa) because of week or no interaction between H_2 gas and adsorbent. The anticipated target adsorption capacity of NCC is still far lower compared to metal nanocomposites (Fig. 12). The study results revealed that the H_2 gas adsorption capacity of metal carbon nanocomposites were significance compared to pure NCC and it shown moderate to excellent hydrogen adsorption in phrase of the proportion matter captured per unit pressure. The Au-doped NCC can guide to a greater H_2 gas depository compare to pure NCC adsorbent.

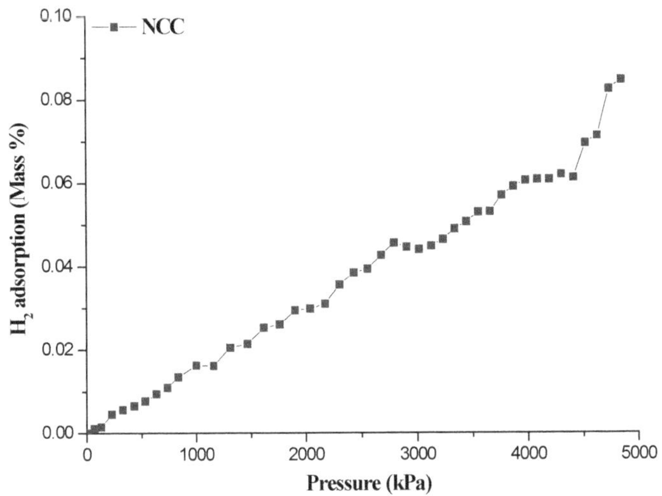

Figure 12. Hydrogen adsorption of metal carbon nanocomposite pristine ***NCC****.*

4.2 Ethylene gas adsorption study

Chemical adsorption of ethylene gas (C_2H_4) is truly significance in agricultural industry during transportation or depository, several harvested-fruits can generate ethylene (C_2H_4) gas which is capable of diminishing the character and self-survival. Lately, many experimental studies of sorption of ethylene (C_2H_4) gas on a range of active *d* block metal composite facet have been done using new techniques. These composite materials help to maintain the freshness of the harvested products in the course of transportation and hoarding. The chemical adsorption of ethylene (C_2H_4) gas was determined on the organic-inorganic metal carbon nanocomposites in a moderate pressure auto gas sorption operation using gas chromatography (GC) at atmospheric temperatures. Typically, 0.50 gm of the five adsorbent materials (**NCC**, **Cu@NCC**, **Ni@NCC**, **K-Mn@NCC**, **Au@NCC**) was placed inside a Pyrex sorption unit joined to the adsorption apparatus and exposed to a regular pre-treatment prior to ethylene gas (C_2H_4) adsorption analysis which integrated vacuum dehydrating in a vacuum desiccator and thermal revival in an oven [27].

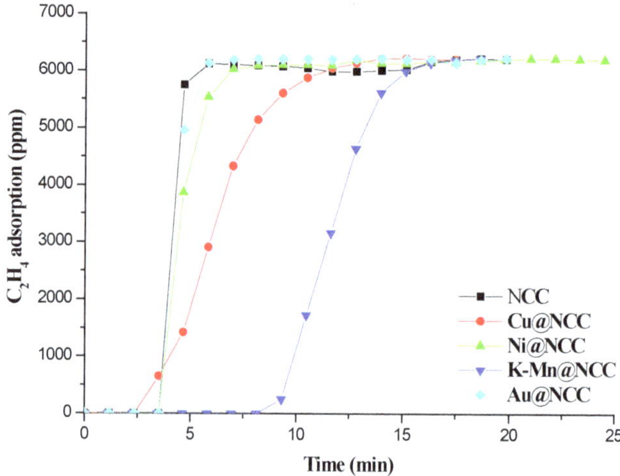

*Figure 13. Ethylene adsorption of carbon nanocomposites **Cu@NCC, Ni@NCC, K-Mn@NCC, Au@NCC** and **NCC**.*

The ethylene (C_2H_4) gas adsorption abilities of the carbon nanocomposites were achieved by gas chromatography (GC) analysis at atmospheric temperature. Fig. 13 depicts the ethylene adsorption visual representation of five carbon nanocomposites. The corresponding C_2H_4 gas adsorptions on the adsorbents (**NCC**, **Cu@NCC**, **Ni@NCC**, **Au@NCC**) have been raised remarkably from 0 to 6000 ppm in 3–15 min while **K-Mn@NCC** displayed a corresponding ethylene (C_2H_4) gas sorption growth starting with 0 to 6000 pm in around 8~15 minutes. The ethylene (C_2H_4) gas sorption was nearly identical for all five of the adsorbents and more or less steady upon sample saturation with time (Fig. 13). The metal carbon nanocomposites had a uniform pore size and a transition-metal attraction which permits the nanocomposites to bind and capture the ethylene alike tinier compounds (3.14 Å) at adsorption sites. These nanocomposites can be broadly used as efficient and low-priced adsorbents in the time of movement and storage of harvested fruits and vegetables in order to protect the farming product quality by capturing ethylene release.

Conclusion

This chapter describes the evidence for the robust research findings carried out in recent time to project a new and improved techniques for the cleansing of water waste from organic dyes along with energy storage applications. The metal carbon nanocomposites **Au@NCC, Cu@NCC, Ni@NCC, K-Mn@NCC, P@NCC, Au-P@NCC, Mo-V@NCC, Pd@NCC, Au-Pd@NCC, W@NCC** and **Au-Salen@NCC** were prepared by a simple pattern method employing low-cost petroleum based pitch as the carbon creator and nano structures silica ball as hole maker. The green, recyclable and sustainable organic-inorganic metal nanocomposites were applied successfully as a solid-sustainable catalyst for degeneration of toxic organic pollutant pigments (Chromotrope-2R, Methylene Blue and Eosin-Y) and as an adsorbent for energy protection (H_2 and C_2H_4) and environment fixing.

References

[1] V.J. Mayani, S.V. Mayani, S.W. Kim, Gold salen complex doped carbon nanocomposite Au(Salen)@CC for catalytic oxidation of Eosin Y and Chromotrope 2R dyes, Nature's Sci. Rep. 7(7239), (2017) 1-9. https://doi.org/10.1038/s41598-017-07707-6

[2] M. Alberta, M.S. Lessina, B. F. Gilchrista, Methylene blue: dangerous dye for neonates, J. Pediatr. Surg., 38(8) (2003), 1244-1245. https://doi.org/10.1016/S0022-3468(03)00278-1

[3] S.V. Mayani, V.J. Mayani, S.W. Kim, Synthesis of molybdovanadophosphoric acid supported hybrid materials and their heterogeneous catalytic activity, Mater. Lett. 111 (2013) 112–115. https://doi.org/10.1016/j.matlet.2013.08.078

[4] I. Fatimah, S. Wang, D. Wulandari, ZnO/montmorillonite for photocatalytic and photochemical degradation of methylene blue. Appl. Clay Sci. 53 (2011) 553–560. https://doi.org/10.1016/j.clay.2011.05.001

[5] P. Chowdhury, S. Athapaththu, A. Elkamel, A.K. Ray, Visible-solar-light-driven photo-reduction and removal of cadmium ion with Eosin Y sensitized TiO_2 in aqueous solution of triethanolamine. Sep. Purif. Technol. 174 (2017) 109–115. https://doi.org/10.1016/j.seppur.2016.10.011

[6] E.M. Arbeloa, C.M. Previtali, S.G. Bertolotti, Photochemical study of eosin-Y with PAMAM dendrimers in aqueous solution. J. Lumin. 180 (2016) 369–375. https://doi.org/10.1016/j.jlumin.2016.08.017

[7] S. Nagaya, H. Nishikiori, H. Mizusaki, K. Sato, H. Wagata, K. Teshima, Crystal structure and photoelectric conversion properties of eosinY adsorbing ZnO films prepared by electroless deposition. Appl. Catal. B. Environ. 189 (2016) 51–55. https://doi.org/10.1016/j.apcatb.2016.02.013

[8] K. Vignesh, A. Suganthi, M. Rajarajan, R. Sakthivadivel, Visible light assisted photo-decolorization of Eosin-Y in aqueous solution using hesperidin modified TiO_2 nanoparticles. Appl. Surf. Sci. 258 (2012) 4592–4600. https://doi.org/10.1016/j.apsusc.2012.01.035

[9] T. Robinson, G. McMullan, R. Marchant, P. Nigam, Remediation of dyes in textile effluent: a critical review on current treatment technologies with a proposed alternative. Bioresourse Technol. 77 (2001) 247–255. https://doi.org/10.1016/S0960-8524(00)00080-8

[10] G.A. Umbuzeiro, H.S. Freeman, S.H. Warren, D.P. Oliveira, Y. Terao, T. Watanabe, L.D. Claxton, The contribution of azo dyes to the mutagenic activity of the Cristais river. Chemosphere 60 (2005), 55–64. https://doi.org/10.1016/j.chemosphere.2004.11.100

[11] K.P. Sharma, S. Sharma, S. Sharma, P.K. Singh, S. Kumar, R. Grover, P.K. Sharma, A comparative study on characterization of textile wastewaters (untreated and treated) toxicity by chemical and biological tests. Chemosphere 69, (2007) 48-54. https://doi.org/10.1016/j.chemosphere.2007.04.086

[12] H. Dong, G. Zeng, L. Tang, C. Fan, C. Zhang, X. He, Y. He, An overview on limitations of TiO_2-based particles for photocatalytic degradation of organic pollutants and the corresponding countermeasures. Water Res. 79 (2015) 128–146. https://doi.org/10.1016/j.watres.2015.04.038

[13] S.V. Mayani, V.J. Mayani, K.G. Bhattacharyya, Phenol and its analogues in water: sources, environmental fate, effects and treatment, In: J. B. Baruah (Eds) In chemistry of phenolic compounds; State of the art, Nova Science Publishers Inc., New York, USA, 2011, Chapter 13, 181–202.

[14] V.J. Mayani, S.V. Mayani, S.W. Kim, Simple preparation of tungsten supported carbon nanoreactors for specific applications: adsorption, catalysis and electrochemical activity. Appl. Surf. Sci. 345, (2015) 433–439. https://doi.org/10.1016/j.apsusc.2015.03.178

[15] V.J. Mayani, S.V. Mayani, Y.G. Lee, S.K. Park, A non-chromatographic method for the separation of highly pure naphthalene crystals from pyrolysis fuel oil. Sep. Purif. Techn. 80 (2011) 90–95. https://doi.org/10.1016/j.seppur.2011.04.013

[16] H. Zhao, N. Neamati, A. Mazumder, S. Sunder, Y. Pommier Jr., T.R. Burke, Arylamide inhibitors of HIV-1 integrase. J. Med. Chem. 40 (1997) 1186–1194. https://doi.org/10.1021/jm960449w

[17] S. Asir, A.S. Demir, H. Icil, The synthesis of novel, unsymmetrically substituted, chiral naphthalene and perylene diimides: photophysical, electrochemical, chiroptical and intramolecular charge transfer properties. Dyes Pigm. 84 (2010) 1–13. https://doi.org/10.1016/j.dyepig.2009.04.014

[18] S.V. Mayani, V.J. Mayani, S.K. Park, S.W. Kim. Synthesis and characterization of metal incorporated composite carbon materials from pyrolysis fuel oil, *Mater. Lett.* 82 (2012) 120–123. https://doi.org/10.1016/j.matlet.2012.05.078

[19] N.G. Asenjo, C. Botas, C. Blanco, R. Santamaría, M. Granda, R. Menéndez, Synthesis of activated carbons by chemical activation of new anthracene oil-based pitches and their optimization by response surface methodology. Fuel Process Technol. 92 (2011) 1987–1992. https://doi.org/10.1016/j.fuproc.2011.05.021

[20] Y.S. Wang, C.Y. Wang, M.M Chen, New mesoporous carbons prepared from pitch by simultaneous templating carbonization. New Carbon Mater. 24 (2009) 187–190. https://doi.org/10.1016/S1872-5805(08)60047-5

[21] I. Mochida, S.H. Yoon, Y. Korai, K. Kanno, Y. Sakai, M. Komatsu, In: H. Marsh F. Rodriguez-Reinoso (Eds) Sciences of carbon materials. Publicaciones de la Universidad de Alicante, London 2000, 259.

[22] S.B. Yoon, J.Y. Kim, J.H. Kim, Y.J. Park, K.R. Yoon, S.K. Park, J.S. Yu, Synthesis of monodisperse spherical silica particles with solid core and mesoporous shell: Mesopore channels perpendicular to the surface. J. Mater. Chem. 17 (2007) 1758-1761. https://doi.org/10.1039/b617471j

[23] V.J. Mayani, S.H.R. Abdi, R.I. Kureshy, N.H. Khan, S. Agrawal, R.V. Jasra, Synthesis and characterization of (S)-amino alcohol modified M41S as effective material for the enantioseparation of racemic compounds. J. Chrom. A, 1135 (2006) 186–193. https://doi.org/10.1016/j.chroma.2006.09.094

[24] V.J. Mayani, S.H. R. Abdi, R.I. Kureshy, N.H. Khan, A. Das, H.C. Bajaj, Heterogeneous Chiral Copper Complexes of Amino Alcohol for Asymmetric Nitroaldol Reaction. J. Org. Chem., 75 (2010) 6191–6195. https://doi.org/10.1021/jo1010679

[25] S.V. Mayani, V.J. Mayani, S.W. Kim, Development of novel porous carbon frameworks through hydrogen-bonding interaction and its ethylene adsorption activity. J. Porous Mater. 19 (2012) 519–527. https://doi.org/10.1007/s10934-012-9565-2

[26] V.J. Mayani, S.V. Mayani, S.W. Kim, Development of nanocarbon gold composite for heterogeneous catalytic oxidation. Mater. Lett. 87 (2012) 90-93. https://doi.org/10.1016/j.matlet.2012.07.071

[27] S.V. Mayani, V.J. Mayani, J.Y. Lee, S.H. Ko, S.K. Lee, S.W. Kim, Preparation of multi metal–carbon nanoreactors for adsorption and catalysis. *Adsorption,* 19 (2013) 251–257. https://doi.org/10.1007/s10450-012-9447-6

[28] V.J. Mayani, S.V. Mayani, S.W. Kim, Development of Gold Phosphorus Supported Carbon Nanocomposites. Bull. Korean Chem. Soc. 35(2) (2014), 401-406. https://doi.org/10.5012/bkcs.2014.35.2.401

[29] V.J. Mayani, S.V. Mayani, S.W. Kim Development of Palladium, Gold and Gold-Palladium Containing Metal-Carbon Nanoreactors: Hydrogen Adsorption. Bull. Korean Chem. Soc. 35(5) (2014), 1312-1316. https://doi.org/10.5012/bkcs.2014.35.5.1312

[30] V.J. Mayani, S.V. Mayani, S.W. Kim, Palladium, Gold, and Gold–Palladium Nanoparticle-Supported Carbon Materials for Cyclohexane Oxidation. Chem. Eng. Comm. 203 (2016) 539–547. https://doi.org/10.1080/00986445.2015.1048800

[31] E.H.E. Safaa, M.E. Mohamed, Degradation of methylene blue by catalytic and photo-catalytic processes catalysed by the organotin-polymer $3\alpha[(Me_3Sn)_4Fe(CN)_6]$. Appl. Catal. B: Environ. 126 (2012) 326–33. https://doi.org/10.1016/j.apcatb.2012.07.032

[32] S.H.R. Abdi, R.I. Kureshy, N.H. Khan, V.J. Mayani, H.C. Bajaj, Dimeric & polymeric chiral Schiff base complexes and supported BINOL complexes as potential recyclable catalysts in asymmetric kinetic resolution and C–C bond formation reactions. Catal. Surv. Asia 13 (2009), 104–131. https://doi.org/10.1007/s10563-009-9071-y

[33] K. Vignesh, A. Suganthi, M. Rajarajan, R. Sakthivadivel, Visible light assisted photo-decolorization of eosin-Y in aqueous solution using hesperidin modified TiO_2 nanoparticles. Appl. Surf. Sci. 258 (2012), 4592–4600. https://doi.org/10.1016/j.apsusc.2012.01.035

[34] L.C. Almeida, S.G. Segura, C. Arias, N. Bocchi, E. Brillas, Electrochemical mineralization of the azo dye Acid Red 29 (Chromotrope 2R) by photoelectro Fenton

 Materials Research Forum LLC
https://doi.org/10.21741/9781644901397-4

process. Chemosphere 89 (2012), 751–758. https://doi.org/10.1016/j.chemosphere.2012.07.007

[35] I.A. Salem, H.A. El-Ghamry, M.A. El-Ghobashy, Application of montmorillonite–Cu(II)ethylenediamine catalyst for the decolorization of Chromotrope 2R with H_2O_2 in aqueous solution. Spectrochim. Acta A: Mol. Biomol. Spectrosc. 139 (2015) 130–137. https://doi.org/10.1016/j.saa.2014.11.053

[36] G.M. Psofogiannakis, T.A. Steriotis, A.B. Bourlinos, E.P. Kouvelos, G.C. Charalambopoulou, A.K. Stubos, G.E. Froudakis, Enhanced hydrogen storage by spill over on metal-doped carbon foam: an experimental and computational study. Nanoscale, 3 (2011) 933–936. https://doi.org/10.1039/c0nr00767f

Advances in Wastewater Treatment II | Materials Research Forum LLC
Materials Research Foundations **102** (2021) 128-167 | https://doi.org/10.21741/9781644901397-5

Chapter 5

A Statistical Approach to Model Selection for Dynamic Adsorption Columns

Paul Musonge*[1,2,a], Manqoba Shezi[1,b]

[1]Institute of Systems Science, Durban University of Technology. Durban, South Africa

[2]Faculty of Engineering, Mangosuthu University of Technology. Durban, South Africa

[a]paulm@dut.ac.za, [b]21314918@dut4life.ac.za

Abstract

A variety of models have been used to describe and predict breakthrough curves for dynamic adsorption systems, in order to scale up laboratory and pilot plant systems. There are however limitations in the applicability of existing models. The study is aimed at providing unambiguous approaches in selecting the best performing model between Thomas, Yoon-Nelson and Bohart-Adams (B-A) models for three dynamic adsorption systems. Three approaches were implemented in this study using published experimental data of three adsorption systems. The first approach was the application of statistical analysis between actual and predicted breakthrough curves without modifying the models. The second and third approaches were application of local mean values (LMV) and global mean values (GMV) of empirical constants to predict breakthrough curves. Predictive and generalization performances of the three models were evaluated using the statistical criteria of Mean Absolute Error (MAE), Root mean Squared Error (RMSE) and Correlation Coefficient (R^2).

Keywords

Acid Mine Drainage, Adsorption, Yoon-Nelson Model, Thomas Model, B-A Model

Contents

Materials Research Foundations **102** (2021) 128-167
https://doi.org/10.21741/9781644901397-5

1. Introduction

Acid mine drainage (AMD) is currently the major pollutant on both land and water in South Africa. The main source of Acid Mine Drainage is oxidation of sulphide mineral ores, which are primarily exposed to the environment by vigorous mining activities [1]. In the 21st century, where environmental protection and rehabilitation are considered as a priority, technical feasibility and economic sustainability are key drivers in determining the most suitable proposed system or method for any application [2]. Biosorption is eco-friendly, cost effective, ease of operation, high efficiency and wide range of applications [3]. Biosorption is the process of binding contaminants (metal ions, dyes etc.) on the surface of the biological material due to the presence of characteristic functional groups e.g. carboxylic, hydroxyl, amino, carbonyl, phosphate, sulphonic etc. [4]. In water treatment by biosorption, the solid phase/substance and pollutants are respectively called adsorbent and adsorbates. The treatment of water by adsorption technology for both batch and column operations depends upon many factors such as pH, nature of adsorbent and adsorbates, temperature, particle size, dose of adsorbent, presence of other pollutants, flow rate, concentration of adsorbates and bed volume/porosity. For a dynamic adsorption system (column adsorption), the method to determine concentration-time profiles (breakthrough curves) is a significant factor since it provides the basic and foremost information for design and performance of a column adsorption system. Breakthrough curve information assist in determining rational scale of the adsorption column for practical applications. Direct experimentation and mathematical modelling are the basic approaches in predicting breakthrough curves. However, mathematical modelling though it is complex for dynamic adsorption processes it is readily realised with minimal experimental work and therefore it has attracted increasing attention in the preceding decades [5]. However, since dynamic adsorption processes operate at unsteady state conditions, the development of a model that predicts concentration-time profiles is laborious in most cases. This is due to the fact that adsorbate(s) concentration in the influent varies as it moves through the bed [6]. Fundamental equations derived to model the dynamic adsorption system with theoretical precision are differential in nature and to solve them, intricate numerical methods are required. Hence, elementary mathematical models such as Thomas, Bohart-Adams, Yoon-Nelson and Dose-Response models have been developed to anticipate the dynamic behaviour of the column [6]. Since technology has dominated the world to an extreme extent, machine learning or artificial intelligence is another alternative path in solving complex problems such as the adsorption process, which depends upon many factors and is operated under unsteady conditions. The artificial neural networks (ANN) are modelling tools that are commonly used to predict complex systems that are difficult to model using conventional modelling methods, such as mathematical modelling [7]. A linear model is not appropriate to constitute a substantial relationship between inputs and outputs for an adsorption process, hence ANN approach is suitable for problems where interactions between variables are nonlinear [8]. From reviewed adsorption systems, a significant observation was that model parameters were observed to vary with operating conditions, hence the response of dynamic models to these variations is not well studied for dynamic adsorption systems. Moreover, the parameter that must be varied to reflect deviations in kinetic behavior is ambiguous for each considered dynamic adsorption model. Typically, refitting is the only alternative technique applied to attain a new set of model parameters when operating conditions vary. However, what if one value of each model parameter can be applied for a certain range of operation in order to mitigate refitting when conditions change? The current work reports the application of critical analyses on model parameters for Fe-sea shell, Pb-charcoal and Pb-*Agirus bisporus* adsorption systems. The critical analyses include applying mean values of model parameters based on range of operation as well as generalized model parameters based on the experimental data in order to predict breakthrough curves for each system. As previously mentioned, biosorption is among one of the complex processes and the interaction between the variables is known to be nonlinear. Thus, three-layered feed-forward backpropagation neural network models for modelling Fe-seashell, Pb-charcoal and Pb-*Agirus bisporus* adsorption systems were developed. The three layers of the networks were input, hidden and output layers. The data for the aforementioned systems was obtained from previously published data. The investigated process variables for considered dynamic adsorption systems were sorbent mass, flow rate, influent sorbate concentration, processing time and bed depth. The efficient operation of considered adsorption systems in plant scale requires optimization of all operating conditions, which involves quite a number of experiments without automation. Hence, artificial neural network models would allow the development and application of automated wastewater treatment plant at low operating costs.

2. Dynamic adsorption systems

Batch mode operation results are very significant when developing column operation; however, the latter have distinct advantages over the former since the rates of adsorption depend upon the concentration of the adsorbate(s) in the solution being treated [3, 9].

2.1 Breakthrough curve characteristics and development

The dynamic behaviour of column operation is best described in terms of concentration-time profile, which is known as breakthrough curve [6]. Typical breakthrough curves are illustrated in Fig. 1. During water purification in a column, a mass transfer zone is developed as illustrated in Fig. 1. The mass transfer zone is the area of the bed where the adsorption process takes place. When the contaminated water starts to flow through the column, the mass transfer zone changes from 0% of the initial concentration which corresponds to the sorbate-free or clean adsorbent to 100% of the initial concentration which corresponds to the complete saturation of the bed [6].

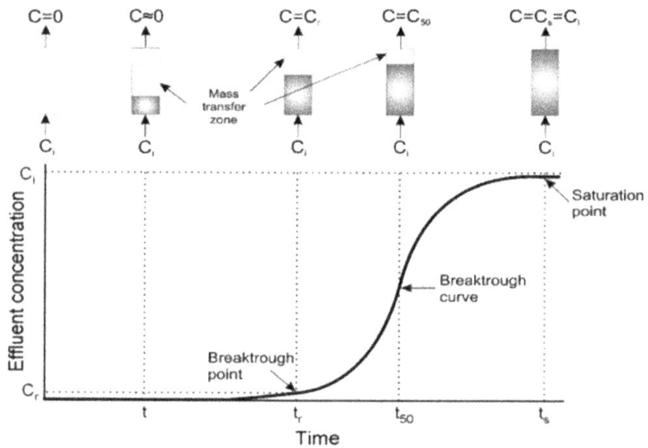

Figure 1. Illustration of a typical breakthrough curve [6].

The breakthrough curve is generated by plotting a graph of nominal concentration, i.e. the ratio of metal (sorbate) concentration in the effluent (C_t) to the influent concentration (C_o) versus time. The nominal concentration falls on the y-axis and processing time on the x-axis. The nominal concentration changes from 0 to 1, and the s-shaped breakthrough curve is generated due to the rapid rise in the effluent concentration. The point at which rapid rise begins is known as the breakthrough point. Column capacity with respect to the influent sorbate concentration and flow rate is highly significant in attainment of breakthrough point. From Fig. 1, breakthrough point is qualitatively indicated. The rapid rise of the breakthrough curve also indicates that the available binding sites of the adsorbent are approaching saturation. Saturation point or exhaustion point indicate a time where the column is totally saturated and sorbate concentrations in the effluent and influent are balanced or equal. However, the most important part of the breakthrough curve as far as metal (sorbate) sorption is concerned comprises only the origin up to breakthrough point section [9]. The breakthrough or service time, t_r or t_b, is generally predetermined by the breakthrough point concentration, C_r or C_b, which is related to the allowable disposal limit

Materials Research Forum LLC
https://doi.org/10.21741/9781644901397-5

for a particular metal (sorbate), which in turn makes it possible to determine volume of water treated [6]. Service time of the packed column is directly reflected by the time required to attain breakthrough point, hence the longer the service time, the higher metal (sorbate) uptake or removal capability of the column. The steepness of the breakthrough curve is the measure of column efficiency to attain the saturation point; however, it is worth noting that the steepness of the breakthrough curve changes with particle size and sorbent type [9].

2.2 Modelling of dynamic adsorption systems

Experimental evaluation of adsorption performance under various conditions is usually high-priced and time-consuming, thus development of mathematical models that have the capability to predict and optimise fixed-bed adsorption is therefore essential.

The unsteady state differential mass balance in the fluid for a section *dz* length of the bed is therefore given by,

$$\varepsilon \frac{\partial C(t,z)}{\partial t} + u \frac{\partial C(t,z)}{\partial z} + (1-\varepsilon)\rho_\rho \frac{\partial q(t,z)}{\partial t} = D_L \frac{\partial^2 C(t,z)}{\partial z^2} \tag{1}$$

Where ε is the porosity of the bed, ρ_p is the sorbent density, other models use the bulk density of the bed, ρ_b which is equal to (1- ε) ρ_p (kg/m^3), u is superficial velocity for an empty bed (m/s), D_L is an axial dispersion coefficient (m^2/s). The first and second terms in Eq. 1 represent accumulation of solute in the liquid and in the solid, respectively. The third term represents the amount of solute flowing in by convention to the section *dz* minus that flowing out of the bed. The last term denotes axial dispersion of the solute in the bed, which results in mixing of the solute and solvent [10].

The initial and boundary conditions for the fixed-bed column initially free of adsorbate and subjected to a step change in the influent sorbate concentration in the column inlet at time zero are given by,

$$\{t = 0, \quad C(t,z) = q(t,z) = 0\}$$
$$\{z = 0, \quad C(0, t = 0) = 0;\ C(0, t > 0) = C_o\}$$
$$\left\{z = Z, \quad \frac{\partial C}{\partial z} = 0\right\}$$

When axial dispersion is ignored, Eq. 1 reduces to;

$$\varepsilon \frac{\partial C(t,z)}{\partial t} + u \frac{\partial C(t,z)}{\partial z} + (1-\varepsilon)\rho_\rho \frac{\partial q(t,z)}{\partial t} = 0 \tag{2}$$

Materials Research Forum LLC
https://doi.org/10.21741/9781644901397-5

Thus, the initial and boundary conditions are then as follows,

$$\{t = 0, \qquad C(t,z) = 0\}$$
$$\left\{z = 0, \qquad \frac{D_L \varepsilon}{u}\frac{\partial C}{\partial z} = C - C_o\right\}$$
$$\left\{z = Z, \qquad \frac{\partial C}{\partial z} = 0\right\}$$

Where t is the time, C_o is the influent sorbate concentration and Z is the bed height. An appropriate rate expression relating the rate of sorbate uptake, $\partial q/\partial t$, to liquid phase concentration, C or solid phase concentration, q or both concentrations is required to complete the model that describes dynamic performance of biosorption.

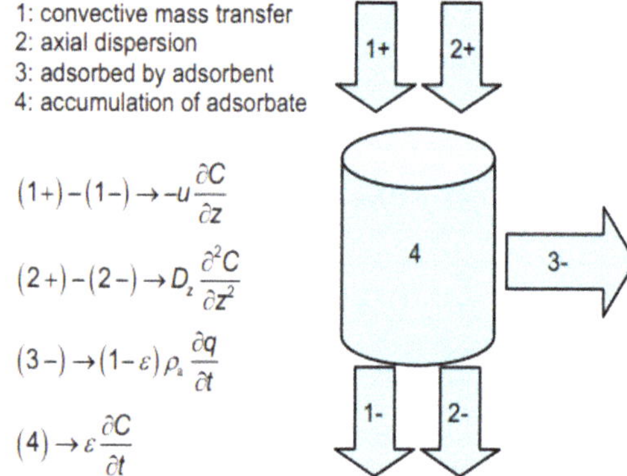

Figure 2. Schematic presentation of mass conservation of a control volume [5].

With reference to Fig. 2 and Eq. 1 (or Eq. 2), in which the former and the latter illustrates schematic presentation and mathematical presentation of mass conservation equation, respectively. The development of the mass balance differential equation is based upon the following assumptions:

- The process operates under isothermal conditions.
- No chemical reaction takes place in the column.
- The sorbent or packing material is made of porous particles that are uniform in size and spherical shape.
- The flow rate is constant and invariant with the column position.
- Each species have activity coefficient that is one.

Materials Research Forum LLC
https://doi.org/10.21741/9781644901397-5

Various mathematical models such as Bohart-Adams, Thomas, Yoon-Nelson models have been developed to predict the dynamic behaviour of the biosorption column and allow some kinetic coefficients to be estimated. Following is the description of these models.

2.2.1 Bohart – Adams (B-A) model and bed depth service time (BDST) model

Bohart-Adams model was developed by [11], when they continued with their work of analysing the typical chlorine charcoal transmission curve. They theorized that "the uptake rate of chlorine is proportional to the concentration of the chlorine existing in the bulk fluid and the residual adsorptive capacity of charcoal". Bohart-Adams model is widely applied in the design of activated carbon adsorbers. Both models assume the rectangular isotherm at equilibrium. The sorbate-sorbent interaction for BDST model is represented by the subsequent quasi-chemical rate differential expression:

$$\frac{\partial q}{\partial t} = k_B C(q_0 - q) \quad (3)$$

As indicated on Fig. 3, the above rate expression shows that at equilibrium ($\partial q/\partial t = 0$). Consequently, Eq. 3, reduces to a rectangular or irreversible relationship between the sorbent and the bulk solution i.e. $q = q_o$.

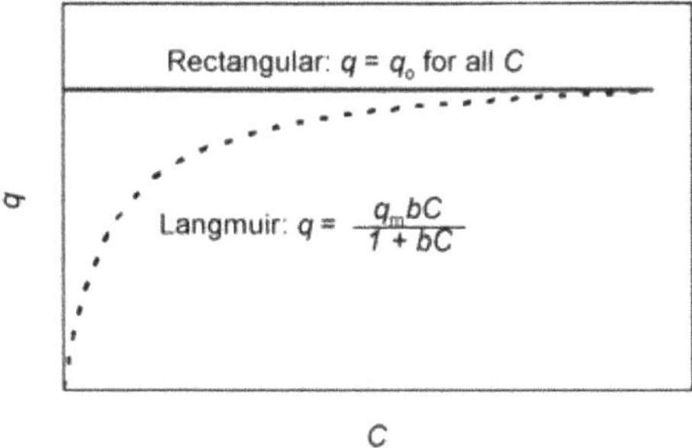

Figure 3. Equilibrium adsorption isotherms illustrating the definition of favourable (Langmuir) and irreversible (rectangular) systems [12].

When axial dispersion is ignored, the analytical solution of equation Eq. 2 and equation Eq. 3 derived by Bohart and Adams is as follows:

Materials Research Foundations **102** (2021) 128-167 | https://doi.org/10.21741/9781644901397-5

$$\frac{C_t}{C_o} = \frac{exp(\alpha)}{exp(\alpha) + exp(\beta) - 1} \quad (4)$$

Where,

$$\beta = \frac{\mathrm{K_B}\rho_\rho \mathrm{q_o z}}{v}\left(\frac{1-\varepsilon}{\varepsilon}\right); \;\; \alpha = \mathrm{K_B} C_o \left(t - \frac{z}{v}\right)$$

In order to simplify the above formal solution to a well-known bed depth service time model, two simplification procedures are considered. Firstly, the two exponential terms exp (α) and exp (β) are frequently much greater than unity, as a result the term "1" on Eq. 4 is neglected. Secondly, sorption time required for the sorbate to exit the column is much greater than the time required for the bulk solution to flow from inlet to outlet of the column, which is represented by z/v. Hence, an assumption can be made that t is much greater than z/v and ignore the z/v term in the expression of α. Finally, the above-mentioned simplifications for equation Eq. 4 yield:

$$\frac{C_o}{C_t} exp(\mathrm{K_B} C_o t) = exp(\mathrm{K_{BA}} C_o t) + exp\left[\frac{\mathrm{K_B}\rho_\rho \mathrm{q_o}(1-\varepsilon)\mathrm{z}}{\varepsilon v}\right] \quad (5)$$

The expression $\rho_\rho \mathrm{q_o}(1-\varepsilon)$ is officially identical to N_o, which is known as the sorption capacity per unit volume of the bed. The product term, εv is equivalent to the superficial velocity u. When replacing $\rho_\rho \mathrm{q_o}(1-\varepsilon)$ and εv in Eq. 5 with N_o and u, respectively and setting nominal concentration (C_t/C_o) as subject of the formula, the well-known BDST model is obtained as shown in Eq. 6:

$$\frac{C_t}{C_o} = \frac{1}{\left[1 + EXP\left(\frac{\mathrm{K_B N_o z}}{u} - \mathrm{K_B C_o t}\right)\right]} \quad (6)$$

In order to derive the well-known B-A model, [5] reviewed vital equations that were based on the hypothesis that Bohart and Adams made during their work of analysing typical chlorine-charcoal transmission curve. The following Eq. 7 and 8 were based on the hypothesis;

$$\frac{\partial C}{\partial z} = -\frac{\mathrm{K_B q_r C}}{u} \quad (7)$$

$$\frac{\partial q_r}{\partial t} = -K_B q_r C \tag{8}$$

Where q_r is the residual adsorptive capacity, additionally the initial conditions were as follows,

$$\left\{t = 0, \quad \frac{q_r}{q_m} = 1\right\}$$

$$\left\{z = 0, \quad \frac{C_t}{C_o} = 1\right\}$$

Consequently, the fundamental form of B-A model is as follows;

$$ln\left(\frac{C_o}{C_t} - 1\right) = ln\left[exp\left(K_B q_m \frac{z}{U_o}\right) - 1\right] - K_B C_o t \tag{9a}$$

The well-known linear of B-A model is given by Eq. 9b below,

$$\ln\left(\frac{C_t}{C_o}\right) = K_{BA} C_o t - K_{BA} N_o \left(\frac{z}{U_o}\right) \tag{9b}$$

Further simplifying Eq. 9b, and setting nominal concentration as subject of the formula, the commonly applied B-A model is obtained, also noting that q_m is equivalent to N_o.

$$\frac{C_t}{C_o} = EXP\left[\left(K_B C_o t - K_B N_o \left(\frac{z}{U_o}\right)\right)\right] \tag{10}$$

Where K_B is the kinetic constant (L/mg.min), N_o is the saturation concentration or sorption capacity per unit volume of fixed bed (mg/L), z is the adsorbent bed depth (cm) and U_o is the linear velocity (cm/min). It is also worth noting that when $exp\left(K_B q_m \frac{z}{U_o}\right)$, in Eq. 9a is much greater than 1, Eq. 10 is obtained an observation that was made by [13]. Empirical constants in both of the models are not constant when the process variable(s) is/are changed which can lead to unsatisfactory predictions of breakthrough curves at 50% breakthrough i.e., nominal concentration (C_t/C_o) of 0.5. The validity of these model is limited to the range of process parameters used according to [14]. Another main limitation of B-A model stated by [15] is that the model is best at predicting initial region of the breakthrough curve at nominal concentration of less than 0.15 but not limited to.

 Materials Research Forum LLC
https://doi.org/10.21741/9781644901397-5

2.2.2 Thomas model

The Thomas model is one of the frequently applied models to estimate the adsorptive capacity of adsorbent and predict breakthrough curves. A reaction model [16] was developed based on the assumptions that axial and radial dispersion in the fixed bed column are negligible and that the pseudo second-order reaction rate principle describes the adsorption process, which reduces to a Langmuir isotherm at equilibrium. The physical properties of the adsorbent (solid-phase) and the liquid phase and column void fraction were also assumed constant. In addition isothermal and isobaric process conditions were assumed and finally negligible intra particle diffusion and external resistance during the mass transfer processes. Hence, theoretically it is suitable to estimate the adsorption process where external and internal diffusion resistances are extremely small. The model consists of seven parameters which are five process variables (influent, effluent concentrations, flow rate, adsorbent mass and processing time) and two empirical constants. The Thomas model predicts better processes that follow Langmuir isotherm, which means that the model is more suitable for monolayer adsorption and homogenous surfaces. The well-known Thomas model is given by:

$$\ln\left(\frac{C_o}{C_t} - 1\right) = \frac{K_{Th} q_o m}{Q} - K_{TH} C_o t \qquad (11)$$

Another significant aspect to consider is that the Thomas model is mathematical equivalent to the BDST model (Eq. 10) since linear velocity or superficial velocity, u or u_o is determined by volumetric flow rate and cross-sectional area:

$$u = U_o = {}^{Q}/_{A_c} \qquad (12)$$

Hence, substituting the above expression into Eq. 10, result in Eq. 13 below,

$$\frac{C_t}{C_o} = \frac{1}{\left[1 + EXP\left(\frac{K_B N_o z A_c}{Q} - K_B C_o t\right)\right]} \qquad (13)$$

Now, when considering that,

$$N_o z A_c = \left(\frac{\text{sorbate mass}}{\text{bed volume}}\right)(\text{bed depth})(\text{bed cross sectional area})$$
$$= \left(\frac{\text{sorbate mass}}{\text{bed volume}}\right)(\text{bed volume}) = \text{sorbate mass}$$

Materials Research Forum LLC
https://doi.org/10.21741/9781644901397-5

And from Eq. 11,

$$q_o m = \left(\frac{\text{sorbate mass}}{\text{sorbent mass}}\right)(\text{sorbent mass}) = \text{sorbate mass}$$

Thus, from the above expressions the following relation can be obtained;

$$N_o z A_c = q_o m \tag{14}$$

Finally, when substituting $q_o m$ in Eq. 6 for $N_o z A_c$ and setting K_B to equal to K_{TH}, Eq. 6 then becomes:

$$\frac{C_t}{C_o} = \frac{1}{\left[EXP\left[\left(\frac{K_{TH} q_o m}{Q} - K_{TH} C_o t\right)\right] + 1\right]} \tag{15}$$

Eq. 15 is clearly identical to Eq. 11 after linearization. As a result of BDST and Thomas models being mathematically equivalent it is therefore not recommended that both of these models are applied in the same study to predict breakthrough curves. In Eq. 15, K_{TH} is the Thomas model constant (L/mg.min), q_o is the adsorption capacity (mg/g), m is the adsorbent mass (g) and Q is the volumetric flow rate (mL/min).

2.2.3 Yoon – Nelson model

In existing adsorption models, the Yoon-Nelson model is extremely concise in form assuming that "the rate of decrease in the probability of adsorption of adsorbate molecule is proportional to the probability of the adsorbate adsorption and the adsorbate breakthrough on the adsorbent" [17]. The model consists of five parameters which are three process variables (influent, effluent concentrations and processing time) and two empirical constants. The rough form of the Yoon-Nelson model makes it less valuable or convenient to obtain process variables and to predict adsorption under variety conditions [5] stated. This model does not consider the effect of each variable during adsorption process. The linearized form of Yoon-Nelson model is as follows;

$$\ln\left(\frac{C_t}{C_o - C_t}\right) = K_{YN} t - K_{YN} \tau \tag{16}$$

Materials Research Forum LLC
https://doi.org/10.21741/9781644901397-5

Where C_t and C_o are effluent and influent adsorbate concentrations (mg/L) respectively, K_{YN} is the rate constant (min^{-1}), t is the processing time (min) and τ is the time required for 50% adsorbate breakthrough (min). The mathematical form of Yoon-Nelson model that is ready to be applied to generate breakthrough curve(s) is given below.

$$\frac{C_t}{C_o} = \frac{EXP(K_{YN}t - K_{YN}\tau)}{[1 + EXP(K_{YN}t - K_{YN}\tau)]} \qquad (17)$$

The conventional method to estimate empirical constants of the dynamic models is through the slopes and intercepts of the linear plots. For B-A model the linear plot is obtained by plotting ln (C_t/C_o) versus t and for BDST and Thomas models the linear plot is obtained by plotting ln ((C_o/C_t)-1) versus t. Finally, linear plot for Yoon-Nelson model is given by ln ((C_t/ (C_o-C_t)) versus t. In this review more emphasis was placed on dynamic adsorption modelling.

3. Factors affecting dynamic adsorption systems

3.1 Adsorbent mass

In predicting concentration-time profiles using modified dynamic models, sorbent mass was varied according to Table 1 below.

Table 1. Effect of sorbent mass at constant flow rate and initial concentration.

Adsorption Systems	m [g]	Q [mL/min]	C [mg/L]
Fe-seashell	30	5	425.4
	60	5	425.4
	90	5	425.4
Pb-charcoal	1.12	15	20
	2.52	15	20
	3.501	15	20
Pb-*Agirus bisporus*	0.4	3	50
	0.8	3	50
	1.2	3	50

Materials Research Foundations **102** (2021) 128-167
https://doi.org/10.21741/9781644901397-5

3.2 Flow rate

The investigated flow rates to predict breakthrough curves were varied according to Table 2 below.

Table 2. Effect of flow rate at constant sorbent mass and initial concentration.

Adsorption Systems	Q	C	m
	[mL/min]	[mg/L]	[g]
Fe-seashell	2.5	425.4	60
	5	425.4	60
	7.5	425.4	60
Pb-charcoal	10	20	2.52
	15	20	2.52
	20	20	2.52
Pb-*Agirus bisporus*	1	50	0.8
	3	50	0.8
	5	50	0.8

3.3 Initial adsorbate concentration

The considered levels of initial concentration under investigation are shown in table below.

Table 3. Effect of initial concentration at constant sorbent and flow rate.

Adsorption Systems	C	m	Q
	[mg/L]	[g]	[mL/min]
Fe-seashell	233.2	60	5
	308.4	60	5
	425.4	60	5
Pb-charcoal	10	2.52	15
	20	2.52	15
	30	2.52	15
Pb-*Agirus bisporus*	20	0.8	3
	50	0.8	3
	100	0.8	3

Materials Research Forum LLC
https://doi.org/10.21741/9781644901397-5

4. Statistical selection approach

4.1 Preliminary quantitative analysis of model performance

When theoretical and experimental breakthrough curves are plotted on the same set of axes, most researchers tend to focus solely on qualitative rather than quantitative analysis. Quantitative analyses are statistical measures of accuracy between the actual and predicted breakthrough curves. The statistical measures do not only assist in generating quantitative conclusions between the actual and predicted breakthrough curves but also assist in determining the overall performance or classification of dynamic models.

Table 4. Summary of performance evaluation metrics in comparing Yoon-Nelson, Thomas and Bohart-Adams models for Fe-seashell system before application of mean values.

Parameters	Yoon – Nelson			Thomas			B – A		
	MAE	RMSE	R^2	MAE	RMSE	R^2	MAE	RMSE	R^2
Mass [g]									
30	0.014	0.021	*0.998*	0.014	0.022	*0.998*	0.255	0.387	*0.530*
60	0.039	0.051	*0.984*	0.040	0.051	*0.984*	0.103	0.133	*0.826*
90	0.044	0.054	*0.979*	0.044	0.053	*0.979*	0.120	0.146	*0.738*
Flow rate [mL/min]									
2.5	0.019	0.024	*0.993*	0.021	0.025	*0.992*	0.087	0.106	*0.839*
5	0.039	0.051	*0.984*	0.040	0.051	*0.984*	0.093	0.120	*0.880*
7.5	0.109	0.170	*0.868*	0.109	0.170	*0.868*	0.133	0.207	*0.551*
Concentration [mg/L]									
233.22	0.106	0.109	*0.908*	0.108	0.111	*0.906*	0.063	0.064	*0.959*
308.4	0.042	0.050	*0.981*	0.042	0.050	*0.981*	0.103	0.123	*0.901*
425.4	0.036	0.046	*0.987*	0.037	0.047	*0.987*	0.325	0.415	*0.666*

Tables 4, 5 and 6 illustrate the actual performance evaluation metrics of the Fe-seashell, Pb-charcoal, and Pb-*Agirus bisporus* adsorption systems preliminary to the application of averaged model parameters. The coefficient of determination, R^2 assesses how good a model predicts output or experimental data. Lower values of RMSE and MAE and higher R^2 signifies a model with better output prediction. The results in Table 4 show that Thomas and Yoon-Nelson models are better at predicting experimental data in contrast to the Bohart-Adams model. The data utilized in generating tables 4, 5 and 6 was obtained from the studies that were done by [18], [19] and [20], respectively.

Table 5. Summary of performance evaluation metrics in comparing Yoon-Nelson, Thomas and Bohart-Adams models for Pb-charcoal system before application of mean values.

Parameters	Yoon – Nelson			Thomas			B – A		
	MAE	RMSE	R^2	MAE	RMSE	R^2	MAE	RMSE	R^2
Mass [g]									
1.12	0.055	0.102	*0.920*	0.055	0.102	*0.920*	0.491	0.913	*0.256*
2.52	0.036	0.090	*0.958*	0.036	0.090	*0.958*	0.572	1.428	*0.322*
3.501	0.015	0.036	*0.993*	0.015	0.036	*0.993*	0.292	0.688	*0.575*
Flow rate [mL/min]									
10	0.020	0.048	*0.987*	0.021	0.052	*0.986*	1.024	2.478	*0.352*
15	0.036	0.091	*0.957*	0.036	0.091	*0.957*	0.573	1.426	*0.321*
20	0.054	0.124	*0.906*	0.055	0.127	*0.902*	0.445	1.027	*0.263*
Concentration [mg/L]									
10	0.045	0.108	*0.934*	0.043	0.103	*0.941*	0.867	2.094	*0.330*
20	0.037	0.092	*0.956*	0.037	0.092	*0.956*	0.570	1.419	*0.320*
30	0.031	0.075	*0.966*	0.032	0.078	*0.963*	0.310	0.746	*0.365*

The performance of the models in Table 5 and 6 for the Pb-charcoal and Pb-*Agirus bisporus* adsorption systems is similar to that of the Fe-seashell adsorption system in Table 4. The Bohart-Adams model continued to result in low R^2 values and higher values of MAE and RMSE, while Yoon-Nelson and Thomas models continued to show close correlations between experimental and predicted data as indicated by higher R^2 values and lower values of error metrics. Tables 4, 5 and 6 show that Yoon-Nelson and Thomas models predict the experimental data similarly when model parameters are not modified to average values. The model validity assessed by employing MAE, RMSE, and R^2 is consistent in all three considered adsorption systems i.e., the Yoon-Nelson and Thomas models outperform the Bohart-Adams model when model parameters are not modified to average values. The Yoon-Nelson and Thomas models are both performing well, however, the Thomas model was selected as the best performing model since the model incorporates significant process variables such as flow rate, mass and initial concentration in contrast to the Yoon-Nelson model. These process variables are very substantial when optimizing dynamic adsorption systems and in scaling up procedures.

Table 6. Summary of performance evaluation metrics in comparing Yoon-Nelson, Thomas and Bohart-Adams models for Pb-Agirus bisporus system before application of mean values.

Parameters	Yoon – Nelson			Thomas			B – A		
	MAE	RMSE	R^2	MAE	RMSE	R^2	MAE	RMSE	R^2
Mass [g]									
0.4	0,031	0,052	*0,987*	0,031	0,051	*0,987*	46,31	77,32	*0,341*
0.8	0,014	0,027	*0,997*	0,013	0,026	*0,997*	67,96	131,4	*0,295*
1.2	0,005	0,011	*0,999*	0,005	0,011	*0,999*	1565	3222	*0.358*
Flow rate [mL/min]									
1	0.006	0.015	*0.998*	0.006	0.014	*0.998*	55.17	133.5	*0.396*
3	0.025	0.048	*0.989*	0.024	0.047	*0.989*	87.50	169.9	*0.284*
5	0.033	0.056	*0.984*	0.034	0.057	*0.983*	264.8	447.9	*0.232*
Concentration [mg/L]									
20	0.011	0.024	*0.996*	0.011	0.024	*0.996*	6.387	13.92	*0.518*
50	0.022	0.042	*0.992*	0.022	0.041	*0.992*	14.47	27.19	*0.359*
100	0.043	0.069	*0.976*	0.043	0.069	*0.976*	882.0	1420	*0.277*

The underperformance of the Bohart-Adams model was expected since the statistical analysis take into account all data points and the Bohart-Adams model is best at predicting the initial part of the breakthrough curve at the nominal concentration of less than 0.15 (Han *et al.* 2009). The observed results are a clear indication that the number of process variables that are included in the model cannot be used to classify the efficacy of the model since the Bohart-Adams model incorporates more process variables than the Thomas model yet the latter outperforms the former. The overall model validity that is preliminary to the application of averaged model parameters is Thomas > Yoon-Nelson > Bohart-Adams. The next section of the study will assess the efficacy of the models when model parameters are modified to average values.

4.2 Application of local mean values (LMV)

This section of the study focuses on the application of local mean values (LMV) of the empirical constants or model parameters to predict breakthrough curves. The local mean values were determined from raw data tables by averaging the model parameters based on the range of operating variables i.e. sorbent mass, flow rate and initial sorbate concentration.

4.2.1 Fe-seashell system

Conventionally, empirical constants of a particular dynamic model vary with operating conditions for a distinct adsorption system. Hence, for this reason, an explicit approach was adopted to investigate how these variations in empirical constants affect model performance.

Table 7. Local mean values (LMV) applied during the prediction of breakthrough curves for Fe-seashell system.

Parameters	Thomas		Yoon-Nelson		B-A	
	K_{TH,ave_1} x 10^{-5} [L/min.mg]	q_{o,ave_1} [mg/g]	K_{YN,ave_1} [min^{-1}]	τ_{ave_1} [min]	K_{B,ave_1} x 10^{-5} [L/min.mg]	N_{o,ave_1} [mg/L]
Mass	±4.62	±20.47	±0.020	±569.9	±1.88	±47129.19
Flow rate	±4.08	±20.8	±0.017	±710.6	±1.80	±44571.34
Concentration	±5.51	±21.9	±0.017	±878.3	±2.58	±38202.06

The local mean values of model parameters for the Fe-seashell adsorption system are tabulated in Table 7. Fig. 4 and 5 demonstrates actual and predicted breakthrough curves at different sorbent masses and flow rates, respectively, whereas Fig. 6 illustrates actual and predicted breakthrough curves at different influent sorbate concentrations. Statistical analysis results and water treatment performance parameters are tabulated in Tables 8 and 9, respectively.

4.2.1.1 Effect of sorbent mass

In predicting concentration-time profiles using modified dynamic models, sorbent mass was varied from 30 to 90 g for the Fe-seashell system. Flow rate and sorbate initial concentration were kept constant at 5 mL/min and 425.4 mg/L, respectively. The sorbent mass was observed to increase with amount of water treated at breakthrough, thus saturation points were reached much quicker at a lower sorbent mass as seen in Fig. 4. Three dynamic models were considered in predicting experimental data namely, Thomas, Yoon-Nelson and Bohart-Adams model.

From Fig. 4 it was observed that the Yoon-Nelson model resulted in practically identical predictions at different sorbent masses when mean values were applied as indicated by one black dashed line. The three predicted breakthrough curves for the Yoon-Nelson model were observed to overlay, thus resulting in identical profiles at different sorbent masses. The results observed for the Yoon-Nelson model signified that no sensitivity was observed as seen in Fig. 4. The Thomas model on the other hand showed close correlation between predicted and actual breakthrough curves at 30 and 60 g, respectively. At 90 g the Thomas model slightly over predicted the experimental data. The observed over prediction at 90 g was attributed to the value of the adsorption capacity (q_o) being further away from the

 Materials Research Forum LLC
https://doi.org/10.21741/9781644901397-5

variable range mean. The sensitivity observed for the Thomas model was attributed to the number of operating parameters incorporated in the model in contrast to the Yoon-Nelson model. The Thomas model was observed to track the experimental data very well for sorbent masses of 30 and 60 g as indicated by R^2 values of 0.999 and 0.984 in Table 8, respectively. It is therefore feasible to apply mean value per variable for a Thomas model to predict breakthrough curves since the predictions were satisfactory and sensitivity was observed. The Bohart-Adams model showed sensitivity when mean values were applied as observed in Fig. 4, however the predictions were not satisfactory. The Bohart-Adams model over predicted the experimental data for all sorbent masses. The observed model validity in Fig. 4 is Thomas > Bohart-Adams > Yoon-Nelson.

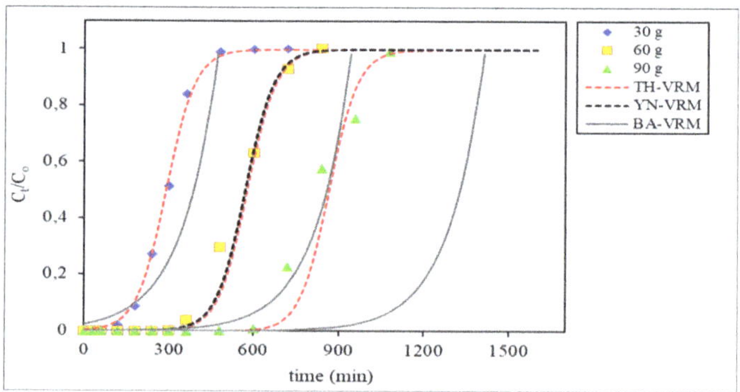

Figure 4. Actual and predicted breakthrough curves for removal of Fe using seashells at different sorbents masses during application of variable range mean values. (TH – Thomas, YN – Yoon-Nelson and BA – Bohart-Adams).

4.2.1.2 Effect of flow rate

The flow rates to predict concentration-time profiles for the Fe-seashell adsorption system were varied from 2.5 to 7.5 mL/min, while the sorbent mass and initial concentration were kept constant at 60 g and 425.4 mg/L, respectively. From Fig. 5, it was observed that flow rate was increasing with a decrease in the amount of water treated at breakthrough. Higher flow rates reduce residence time, thus increasing the saturation rate of the bed.

Materials Research Forum LLC
https://doi.org/10.21741/9781644901397-5

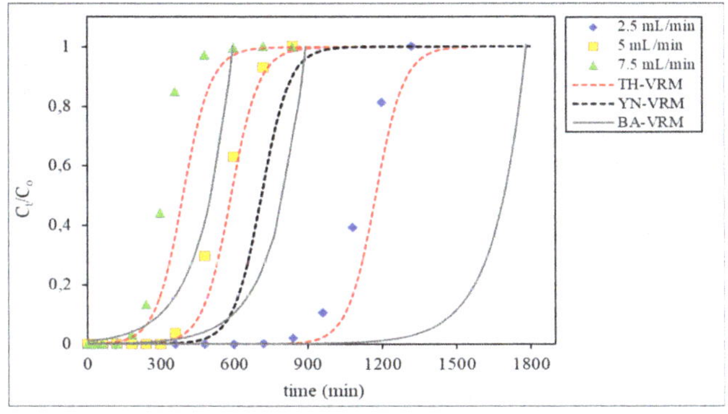

Figure 5. Actual and predicted breakthrough curves for removal of Fe using seashells at different flow rates during application of variable range mean. (TH – Thomas, YN – Yoon-Nelson and BA – Bohart-Adams).

Comparable nature of the curves was observed for the Yoon-Nelson model in Fig. 5. The observed results for the Yoon-Nelson model were analogous to those that were seen in Fig. 4. The Yoon-Nelson model resulted in profiles that had identical profiles when variable range mean values were applied. The Thomas model was seen closer to experimental breakthrough curves when compared to the Yoon-Nelson and Bohart-Adams models. The Thomas model was in line with the experimental data at breakthrough point ($C_t/C_o = 0.05$) for the flow rates of 5 and 7.5 mL/min. At 2.5 mL/min the Thomas model was seen over predicting the experimental data. The observed over prediction at 2.5 mL/min was attributed to the value of the kinetic constant (K_{TH}) being lower than the variable range mean. The standard deviation was lower for K_{TH} and higher for q_o for the results in Fig. 4, contrasting with the standard deviation results for the profiles in Fig. 5, thus the shift of the breakthrough curve is highly influenced by the kinetic constant (K_{TH}) for the Thomas model. The Thomas model was observed to track the experimental data very well for the flow rate of 5 mL/min as indicated by R^2 value of 0.985 tabulated in Table 8. Process parameters such as flow rate, sorbent mass and initial concentration of sorbate play a pivotal role in ensuring higher predictive capabilities for the Thomas model. The Bohart-Adams model on the other hand was observed to overpredict the experimental data for all flow rates studied when variable range mean values were applied as seen in Fig. 5. The profiles in Fig. 5 for the Bohart-Adams model were significantly worse than those observed in Fig. 4. It is worth emphasizing that the Bohart-Adams model profiles were unexpected since the model incorporated most operating variables that are significant for a dynamic

Materials Research Forum LLC
https://doi.org/10.21741/9781644901397-5

adsorption system. The exclusion of significant process parameters such as flow rate and sorbent mass in the Yoon-Nelson model makes it virtually impossible for the model to be applied in real life situations where process parameter(s) may vary. Thus, it is not feasible to apply Bohart-Adams and Yoon-Nelson models in practical applications when considering fixing the model parameters.

4.2.1.3 Effect of initial concentration

The levels of initial sorbate concentration under investigation were 233.22, 308.4 and 425.4 mg/L, while sorbent mass and flow rate were kept constant at 60 g and 5 mL/min for the Fe-seashell system. Figure 6 show that the initial sorbate concentration is decreasing with an increase in the amount of water treated at the breakthrough point. The saturation point was reached much quicker at a higher initial sorbate concentration as seen in figure 6.

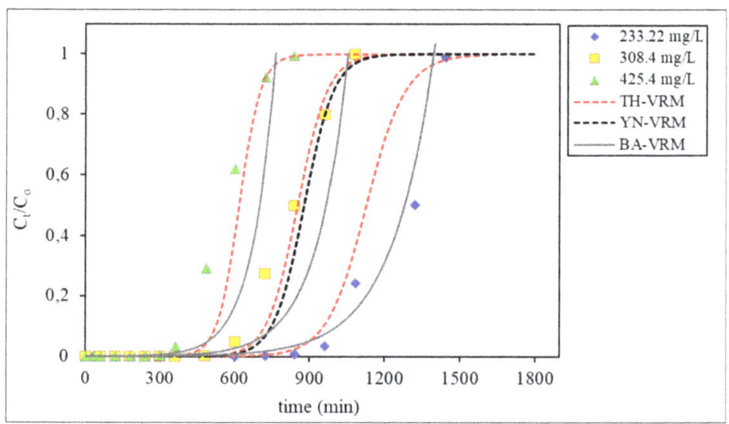

Figure 6. Actual and predicted breakthrough curves for removal of Fe using seashells at different initial concentrations during application of variable range mean (TH – Thomas, YN – Yoon-Nelson and BA – Bohart-Adams).

When variable mean values for model parameters were applied, Yoon-Nelson model resulted in comparable concentration-time profiles at different sorbate concentrations as seen in Fig. 6. The nature of the profiles observed for the Yoon-Nelson model was similar for all cases studied as seen in Fig. 4, 5 and 6. The Yoon-Nelson model in its current form will always require experimental work when process parameter(s) vary which may take time to complete and may delay the adsorption process. Thus, practical application of this is very limited. On the contrary, the Thomas and Bohart-Adams models were seen over predicting and under predicting the experimental data at breakthrough point, with the former being the most dominant as seen in Fig. 6. From the profiles for sorbate concentrations of 308.4 and 425.4 mg/L, the models were seen overpredicting the experimental data when considering the breakthrough section in Fig. 6. At 233.22 mg/L, the models resulted in profiles that were under predicting the experimental data at breakthrough. The overall performance of the Thomas model was consistent for all cases studied, however the standard deviation values of the model parameters that were used to predict profiles in Fig. 6 were above 1. The Thomas model, nonetheless, continued to outperform the Yoon-Nelson and Bohart-Adams models. The Bohart-Adams model in Fig. 6 continued to result in profiles that were deviating from the ideal s-shape, similar to those that were seen in Fig. 4 and 5. The exclusion of intra-particle diffusion in the Bohart-Adams model has a greater impact on the performance of this model, though some authors argue that the Bohart-Adams model is best in predicting the initial part of breakthrough curve before 50 % breakthrough point, supporting the nature of the curves observed on Fig. 6. It is worth noting that although the ideal shape of the breakthrough curve is s-shaped, the breakthrough point is the most prominent part of the curve as it determines the amount of water treated before saturation is attained. Consequently, the aforementioned fact support the nature of the profiles observed for the Bohart-Adams model.

Table 8. Summary of performance evaluation metrics in comparing Yoon-Nelson, Thomas and Bohart-Adams models for the Fe-seashell system during application of local mean values (LMV) of empirical constants.

Parameters	Yoon – Nelson			Thomas			B – A		
	MAE	RMSE	R^2	MAE	RMSE	R^2	MAE	RMSE	R^2
Mass [g]									
30	0.186	0.283	*0.528*	0.011	0.017	*0.999*	3.191	4.843	*0.317*
60	0.037	0.047	*0.987*	0.040	0.051	*0.984*	0.054	0.070	*0.732*
90	0.214	0.260	*0.702*	0.066	0.080	*0.948*	0.007	0.009	*0.809*
Flow rate [mL/min]									
2.5	0.277	0.338	*0.492*	0.044	0.054	*0.962*	0.002	0.003	*0.878*
5	0.101	0.131	*0.813*	0.038	0.049	*0.985*	0.081	0.104	*0.747*
7.5	0.157	0.245	*0.343*	0.096	0.150	*0.896*	1.056	1.646	*0.340*
Concentration [mg/L]									
233.22	0.297	0.304	*0.567*	0.112	0.115	*0.904*	0.052	0.053	*0.983*
308.4	0.057	0.069	*0.961*	0.047	0.057	*0.976*	0.128	0.154	*0.822*
425.4	0.054	0.069	*0.547*	0.071	0.091	*0.947*	0.327	0.417	*0.633*

The results in Table 8 show that the Thomas model is better at predicting the experimental data in contrast to the other two models. The Yoon-Nelson and Bohart-Adams models resulted in low R^2 values and higher values of MAE and RMSE, while the Thomas model showed close correlations between experimental and predicted data as indicated by higher R^2 values and lower values of error metrics. Application of variable range mean values changed the overall model validity from Thomas > Yoon-Nelson > Bohart-Adams to Thomas > Bohart-Adams > Yoon-Nelson. Thomas model was the best performing model for this approach, followed by the Bohart-Adams model.

Table 9. Summary of water treatment performance parameters in comparing Yoon-Nelson, Thomas and Bohart-Adams models for the Fe-seashell system during application of variable range mean (VRM) of empirical constants.

Parameters	Experimental		Yoon – Nelson		Thomas		B – A	
	V_b [L]	q_b [mg]	V_b [L]	q_b [mg]	V_b [L]	q_b [mg]	V_b [L]	q_b [mg]
Mass [g]								
30	*0.745*	301.1	2.1	848.7	*0.695*	280.9	*0.485*	196.0
60	*1.84*	743.6	2.1	848.7	*2.14*	864.8	*2.845*	1150
90	*3.125*	1263	2.1	848.7	*3.585*	1449	*5.205*	2103
Flow rate [mL/min]								
2.5	*2.2*	889.1	1.353	546.6	*2.51*	1014	*3.485*	1408
5	*1.825*	737.5	2.705	1093	*2.085*	842.6	*2.51*	1014
7.5	*1.425*	575.9	4.058	1640	*1.665*	672.9	*1.538*	621.3
Concentration [mg/L]								
233.22	*4.85*	1075	3.525	781.0	*4.49*	994.8	*4.48*	992.6
308.4	*3.025*	886.3	3.525	1033	*3.395*	994.7	*3.39*	993.2
425.4	*1.85*	747.6	3.525	1425	*2.46*	994.2	*2.46*	994.2

To further classify the dynamic models, performance parameters such as the volume of water treated at breakthrough point (V_b) and total mass of metal removed at breakthrough point (q_b) were also evaluated and compared to experimental performance parameters. These performance or design parameters are very significant when assessing the efficacy of dynamic models at the breakthrough point section. Table 9 show experimental and predicted performance parameters after the application of variable range mean. The results in Table 9 show that the Thomas model tracks the experimental data very well, in contrast to the Yoon-Nelson and Bohart-Adams models. The observed model validity in Table 9 is Thomas > Bohart-Adams > Yoon-Nelson.

4.2.2 Pb-Charcoal System

An unambiguous approach of utilizing mean values per variable (or variable range mean) was also applied in the Pb-charcoal system to predict concentration-time profiles. Mean values of model parameters and statistical analysis results for the Pb-charcoal system are tabulated in Tables 10 and 11, respectively.

Table 10. Local mean values (LMV) applied during the prediction of breakthrough curves for Pb-charcoal system.

Parameters	Thomas		Yoon-Nelson		B-A	
	K_{TH,ave_1} x 10^{-5} [L/min.mg]	q_{o,ave_1} [mg/g]	K_{YN,ave_1} [min^{-1}]	τ_{ave_1} [min]	K_{B,ave_1} x 10^{-5} [L/min.mg]	N_{o,ave_1} [mg/L]
Mass	±71.3	±64.7	±0.0143	±479.9	±36.5	±11795.2
Flow rate	±77.7	±55.9	±0.0155	±497.8	±47.2	±8554.7
Concentration	±92.4	±54.6	±0.0145	±501.5	±60.7	±8593.2

Table 11. Summary of performance evaluation metrics in comparing Yoon-Nelson, Thomas and Bohart-Adams models for Pb-charcoal system during application of Local mean values (LMV) of empirical constants.

Parameters	Yoon – Nelson			Thomas			B – A		
	MAE	RMSE	R^2	MAE	RMSE	R^2	MAE	RMSE	R^2
Mass [g]									
1.12	0.152	0.282	*0.578*	0.020	0.038	*0.988*	7.555	14.06	*0.161*
2.52	0.024	0.059	*0.982*	0.053	0.131	*0.908*	0.061	0.153	*0.376*
3.501	0.068	0.160	*0.866*	0.055	0.130	*0.850*	0.003	0.006	*0.646*
Flow rate [mL/min]									
10	0.055	0.134	*0.908*	0.047	0.114	*0.909*	0.021	0.051	*0.516*
15	0.033	0.082	*0.965*	0.019	0.047	*0.989*	0.756	1.882	*0.288*
20	0.117	0.271	*0.627*	0.052	0.120	*0.920*	4.984	11.50	*0.146*
Concentration [mg/L]									
10	0.026	0.063	*0.980*	0.045	0.108	*0.601*	0.003	0.008	*0.607*
20	0.035	0.086	*0.961*	0.015	0.036	*0.994*	1.349	3.360	*0.216*
30	0.087	0.211	*0.768*	0.024	0.058	*0.983*	469.1	1130.6	*0.087*

Local mean values approach proved to be very pertinent for dynamic models under evaluation, particularly for the Thomas model. The results in Table 11 show that the Bohart-Adams model yield low regression values and higher values of MAE and RMSE. The Thomas model on the other hand show a close relation between experimental and theoretical breakthrough curves as indicated by higher regression values and lower values of MAE and RMSE. The Yoon-Nelson model did not mimic changes in operating conditions when mean values were applied, consequently, model validity assessed by employing two error metrics and coefficient of determination was observed to be Thomas

> Bohart-Adams > Yoon-Nelson. The Thomas model continued to outperform the Yoon-Nelson and Bohart-Adams models.

Table 12. Summary of water treatment performance parameters in comparing Yoon-Nelson, Thomas and Bohart-Adams models for the Pb-charcoal system during application of variable range mean (VRM) of empirical constants.

Parameters	Experimental		Yoon – Nelson		Thomas		B – A	
	V_b [L]	q_b [mg]	V_b [L]	q_b [mg]	V_b [L]	q_b [mg]	V_b [L]	q_b [mg]
Mass [g]								
1.12	*0.975*	18.53	4.11	78.09	*0.54*	10.26	*0*	0
2.52	*4.62*	87.78	4.11	78.09	*5.07*	96.33	*8.79*	167.0
3.501	*6.75*	128.3	4.11	78.09	*8.235*	156.5	*14.78*	280.7
Flow rate [mL/min]								
10	*4.5*	85.5	3.09	58.71	*5.15*	97.85	*7.67*	145.7
15	*4.62*	87.8	4.635	88.07	*4.2*	79.8	*6.075*	115.4
20	*3*	57	6.18	117.4	*3.26*	61.94	*4.5*	85.5
Concentration [mg/L]								
10	*6.375*	60.56	4.485	42.61	*9*	85.5	*14.39*	136.7
20	*4,5*	85,5	4.485	85.22	*4.5*	85.5	*7.2*	136.8
30	*2,85*	81,2	4.485	127.8	*3*	85.5	*4.8*	136.8

Table 12 show experimental and predicted performance parameters after the application of variable range mean values for the Pb-charcoal system. The results observed show that the Thomas model tracks the experimental data very well, in contrast to the Yoon-Nelson and Bohart-Adams models. The observed model validity in Table 12 is Thomas > Bohart-Adams > Yoon-Nelson.

4.2.3 Pb-Agirus bisporus system

The approach of applying local mean values (LMV) of model parameters was also adopted in the Pb-*Agirus bisporus* system. Mean values of model parameters are tabulated in Table 13. The statistical analysis and water treatment parameters results for the aforementioned system are tabulated in Tables 14 and 15, respectively.

Materials Research Forum LLC
https://doi.org/10.21741/9781644901397-5

Table 13. Local mean values (LMV) applied during the prediction of breakthrough curves for Pb-Agirus bisporus system.

Parameters	Thomas		Yoon-Nelson		B-A	
	K_{TH,ave_1} x 10^{-5} [L/min.mg]	q_{o,ave_1} [mg/g]	K_{YN,ave_1} [min^{-1}]	τ_{ave_1} [min]	K_{B,ave_1} x 10^{-5} [L/min.mg]	N_{o,ave_1} [mg/L]
Mass	±62,5	±63,3	±0,032	±330	±55,6	±6829,30
Flow rate	±62,1	±63,6	±0,031	±533,2	±56,3	±6814,39
Concentration	±62,6	±63,2	±0,030	±429,8	±55,8	±6787,03

Table 14. Summary of performance evaluation metrics in comparing Yoon-Nelson, Thomas and Bohart-Adams models for Pb-Agirus bisporus system during application of Local mean values (LMV) of empirical constants.

Parameters	Yoon – Nelson			Thomas			B – A		
	MAE	RMSE	R^2	MAE	RMSE	R^2	MAE	RMSE	R^2
Mass [g]									
0.4	0.067	0.112	*0.524*	0.048	0.080	*0.962*	15.32	25.58	*0.441*
0.8	0.022	0.042	*0.992*	0.012	0.024	*0.997*	134.7	260.4	*0.267*
1.2	0.129	0.266	*0.681*	0.034	0.070	*0.974*	15537	31984	*0.280*
Flow rate [mL/min]									
1	0.172	0.416	*0.232*	0.039	0.094	*0.947*	1769	4283	*0.290*
3	0.098	0.189	*0.445*	0.022	0.042	*0.992*	199.2	386.8	*0.253*
5	0.001	0.002	*0.293*	0.021	0.036	*0.992*	26.99	45.65	*0.323*
Concentration [mg/L]									
20	0.171	0.373	*0.375*	0.020	0.044	*0.980*	0.783	1.707	*0.653*
50	0.088	0.165	*0.839*	0.022	0.041	*0.992*	169.8	319.2	*0.274*
100	0.007	0.012	*0.437*	0.092	0.147	*0.899*	2447	3940	*0.253*

The results in Table 14 show that the Thomas model is better at predicting the experimental data in contrast to the other two models. The Yoon-Nelson and Bohart-Adams models resulted in low R^2 values and higher values of MAE and RMSE, while the Thomas model showed close correlations between experimental and predicted data as indicated by higher R^2 values and lower values of error metrics. Model validity assessed by employing two error metrics and coefficient of determination was observed to be Thomas > Bohart-Adams > Yoon-Nelson. The Thomas model continued to outperform the Yoon-Nelson and Bohart-Adams models.

Table 15. Summary of water treatment performance parameters for the Yoon-Nelson, Thomas and Bohart-Adams models for the Pb-Agirus bisporus system during application of Local mean values (LMV) of empirical constants.

Parameters	Experimental		Yoon – Nelson		Thomas		B – A	
	V_b [L]	q_b [mg]	V_b [L]	q_b [mg]	V_b [L]	q_b [mg]	V_b [L]	q_b [mg]
Mass [g]								
0.4	*0,348*	16,5	0,711	33,8	*0,225*	10,7	*0,162*	7,7
0.8	*0,735*	34,9	0,711	33,8	*0,732*	34,8	*0,642*	30
1.2	*1,041*	49	0,711	33,8	*1,239*	59	*0,546*	26
Flow rate [mL/min]								
1	*0,94*	44,7	0,439	20,9	*0,924*	44	*0,857*	41
3	*0,72*	34,2	1,317	63	*0,735*	34,9	*0,645*	31
5	*0,7*	33,3	2,195	104	*0,545*	25,9	*0,430*	20,4
Concentration [mg/L]								
20	*1,8*	34	0,999	19,0	*1,824*	34,7	*1,596*	30,3
50	*0,735*	34,9	0,999	47	*0,729*	34,6	*0,639*	30,4
100	*0,39*	37,1	0,999	95	*0,366*	34,8	*0,318*	30,2

Table 15 show experimental and predicted performance parameters after the application of Local mean values (LMV) for the Pb-*Agirus bisporus* system. The results observed show that the Thomas model tracks the experimental data very well, in contrast to the Yoon-Nelson and Bohart-Adams models. The Yoon-Nelson model did not mimic changes in operating conditions when mean values were applied as indicated by constant values of V_b when sorbent mass and initial concentration were varied, furthermore, the change was only observed at varying flow rates but at constant breakthrough time (t_b). The observed model validity in Table 15 is Thomas > Bohart-Adams > Yoon-Nelson.

4.3 Application of global mean values (GMV)

The mean values used in this section of the study were determined based on the entire experimental domain in order to generalize the conditions, as a result, a single value of the mean for each model parameter was obtained in order to predict breakthrough curves. The averaged model parameters are tabulated in tables 16, 19 and 22 for the three considered systems, respectively. The idea behind the chosen approach was investigating the possibility of fixing model parameter(s) to predict dynamic performance of an adsorption column under varying operating conditions.

4.3.1 Fe-seashell system

How the existing dynamic models respond to changes in operating conditions is not well studied for dynamic adsorption systems, moreover, the parameter that must be varied to

Materials Research Forum LLC
https://doi.org/10.21741/9781644901397-5

reflect deviations in kinetic behaviour is ambiguous for a particular dynamic model. Typically, refitting is the only alternative technique used to obtain new set of model parameters when operating conditions vary. A generalization approach was applied on the Fe-seashell adsorption system to predict breakthrough curves. The approach evaluated model performance under generalized conditions and the metric performance of each model was evaluated and tabulated in Table 17. The water treatment performance parameters are tabulated in Table 18.

Table 16. Global mean values (GMV) of empirical constants applied during the prediction of breakthrough curves for Fe-seashell system.

Thomas		Yoon-Nelson		B-A	
K_{TH,ave_2} x 10^{-5} [L/min.mg]	q_{o,ave_2} [mg/g]	K_{YN,ave_2} [min^{-1}]	τ_{ave_2} [min]	K_{B,ave_2} x 10^{-5} [L/min.mg]	N_{o,ave_2} [mg/L]
±4.74	±21.06	±0.018	±719.58	±2.09	±43300.87

Fig. 7 and 8 illustrate actual and predicted breakthrough curves at different sorbent masses and flow rates for all models under scrutiny, respectively. Fig. 9 on the other hand demonstrate the actual and predicted breakthrough curves at different influent sorbate concentrations. Generalizing empirical constants for the Yoon-Nelson model resulted in comparable predictions despite changing operating conditions, as can be seen in Fig. 7, 8 and 9. The behaviour of the Yoon-Nelson model can be seen from its mathematical form prior to fixing the empirical constants. When predicting dynamic performance of an adsorption column, the Yoon-Nelson model reduces to the form that does not contain any significant operating variable, hence the model failed to predict experimental data when the empirical constants were fixed.

4.3.1.1 Effect of sorbent mass

The first operating variable used to investigate the dynamic performance of an adsorption column was sorbent mass which was varied from 30 to 90 g. Furthermore, in predicting concentration-time profiles using generalized dynamic models for varying sorbent mass, the flow rate and initial sorbate concentration were kept constant at 5 mL/min and 425.4 mg/L, respectively. The generalized Thomas model predictions were observed to be sensitive to generalized mean values as shown in Fig. 7. Close correlation between predicted and actual breakthrough curves was observed for the generalized Thomas model.

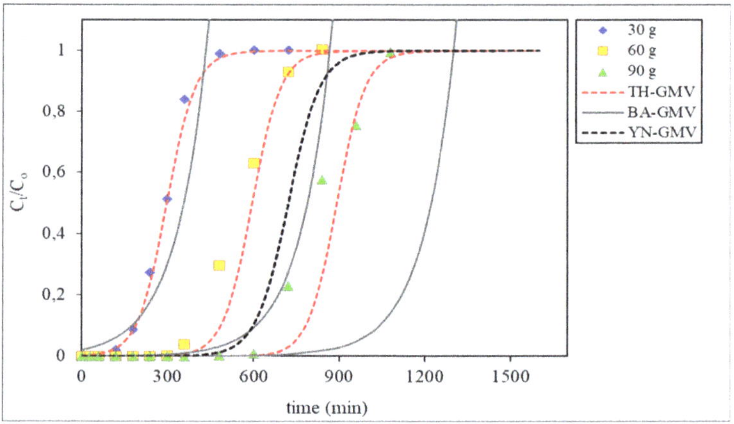

Figure 7. Actual and predicted breakthrough curves for removal of Fe using seashells at different sorbents masses during application of generalized empirical constants (TH – Thomas, YN – Yoon-Nelson and BA – Bohart-Adams).

Likewise, the generalized Bohart-Adams model predictions showed sensitivity when generalized mean values were applied, however, poor correlation between actual and predicted breakthrough curves was observed in Fig. 7 for the generalized Bohart-Adams model. The generalized Bohart-Adams model overpredicted the actual data at sorbent masses of 60 and 90 g, respectively. The generalized Thomas model on the other hand was closely correlated with experimental data at a sorbent mass of 30 g as shown by R^2 value of 0.998, conversely, the generalized Thomas model overpredicted for 60 and 90 g at breakthrough point but R^2 values remained above 0.9. The R^2 values observed for the generalized Bohart-Adams model were increasing from 0.2 to 0.7 when the sorbent mass was varied from 30 to 90 g.

4.3.1.2 Effect of flow rate

The general approach continued to reveal that Yoon-Nelson model does not depend on operating conditions when empirical constants were generalized as seen in Fig. 8. As a result, for the Yoon-Nelson model refitting will always be a requirement whenever operating conditions vary. The predictions obtained using the generalized Thomas and Bohart-Adams models continued to show sensitivity, however with maximal deviations

Materials Research Forum LLC
https://doi.org/10.21741/9781644901397-5

from actual breakthrough curves for the Bohart-Adams model in contrast to the Thomas model.

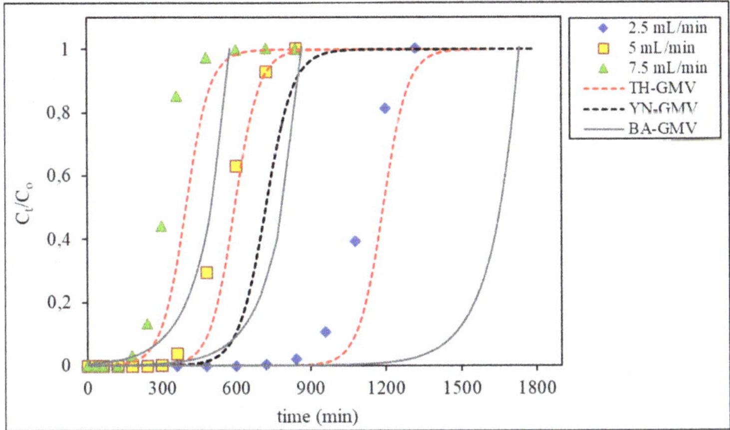

Figure 8. Actual and predicted breakthrough curves for removal of Fe using seashells at different flow rates during application of generalized empirical constants (TH – Thomas, YN – Yoon-Nelson and BA – Bohart-Adams).

All predictions of the generalized Bohart-Adams model were non-ideal for a dynamic sorption system, despite the observed sensitivity. Fig. 8 also showed that the generalized Thomas and Bohart-Adams models were over predicting the experimental data most significantly for the flow rate of 7.5 mL/min. The observed over prediction of the models was attributed to individual model parameters deviating from the generalized mean values. The observed R^2 values of the Thomas model were above 0.9 at the flow rates of 2.5 and 5 mL/min, however at 7.5 mL/min the R^2 obtained was 0.87. The observed R^2 values for the Bohart-Adams model decreased from 0.8 to 0.3 when the flow rate was varied from 2.5 to 7.5 mL/min.

4.3.1.3 Effect of initial concentration

The concentration-time profiles demonstrating the effects of influent concentration on dynamic performance of an adsorption column under generalized conditions for the Fe-seashell system are shown in Fig. 9. The profiles represented the efficacy of the generalized Thomas, Bohart-Adams and Yoon-Nelson models. The levels of initial sorbate concentration under investigation were 233.22, 308.4 and 425.4 mg/L, while the sorbent mass and flow rate were fixed at 60 g and 5 mL/min, respectively.

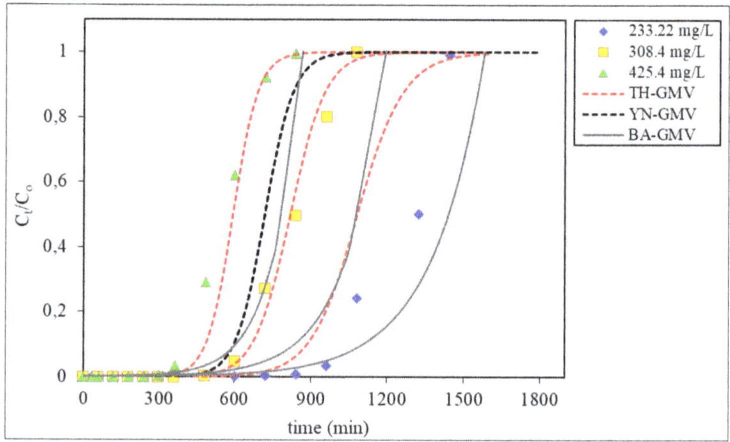

Figure 9. Actual and predicted breakthrough curves for removal of Fe using seashells at different initial concentrations during application of generalized empirical constants (TH – Thomas, YN – Yoon-Nelson and BA – Bohart-Adams).

From Fig. 9, it was observed that the generalized Thomas model outperformed the other two models in predicting concentration-time profiles. The generalized Thomas model was observed to under predict for the concentration of 233.2 mg/L, as a result, the breakthrough time was seen at 817 min versus 970 min which corresponded to 4.085 L versus 4.85 L of water treated at breakthrough (C_t/C_o = 0.05).

From Fig. 9, it is observed that the generalized Thomas model was closely correlated with experimental data at a sorbate concentrations of 308.4 and 425.2 mg/L. The individual model parameters for the Thomas model were close to the generalized mean values, thus explaining the observed superior performance at 308.4 and 425.2 mg/L. The R^2 value was observed to be above 0.9 for a sorbate concentrations of 308.4 and 425.2 mg/L as tabulated in Table 17. At 233.2 mg/L the R^2 obtained was 0.87 for the generalized Thomas model. The Bohart-Adams model on the other hand over predicted the experimental data for all concentrations studied, furthermore, the R^2 value decreased from 0.9 to 0.7 when the concentration was increased from 233.2 to 425.2 mg/L.

Table 17. Summary of performance evaluation metrics in comparing Yoon-Nelson, Thomas and Bohart-Adams models for Fe-seashell system during application of the Global means of the empirical constants.

Parameters	Yoon – Nelson			Thomas			B – A		
	MAE	RMSE	R^2	MAE	RMSE	R^2	MAE	RMSE	R^2
Mass [g]									
30	0.162	0.245	*0.313*	0.013	0.019	*0.998*	6.291	9.549	*0.289*
60	0.106	0.136	*0.788*	0.050	0.064	*0.974*	0.104	0.134	*0.695*
90	0.086	0.105	*0.939*	0.080	0.097	*0.916*	0.016	0.019	*0.776*
Flow rate [mL/min]									
2.5	0.275	0.335	*0.500*	0.059	0.072	*0.930*	0.002	0.003	*0.837*
5	0.106	0.136	*0.788*	0.050	0.064	*0.974*	0.104	0.134	*0.695*
7.5	0.156	0.243	*0.325*	0.107	0.166	*0.876*	1.688	2.630	*0.299*
Concentration [mg/L]									
233.22	0.387	0.396	*0.366*	0.130	0.133	*0.875*	0.017	0.017	*0.988*
308.4	0.099	0.118	*0.927*	0.034	0.040	*0.989*	0.041	0.049	*0.884*
425.4	0.105	0.133	*0.799*	0.047	0.060	*0.978*	0.102	0.130	*0.708*

The statistical analysis results in Table 17 show that the Bohart-Adams model yield low regression values and higher values of MAE and RMSE. The Thomas model on the other hand show a close relation between experimental and theoretical breakthrough curves as indicated by higher regression values and lower values of MAE and RMSE. The Yoon-Nelson model did not mimic changes in operating conditions when mean values were applied, thus, model validity assessed by employing two error metrics and coefficient of determination was observed to be Thomas > Bohart-Adams > Yoon-Nelson.

Table 18 show experimental and predicted performance parameters after the application of *Global means* for the Fe-seashell system. The results observed show that the Thomas model tracks the experimental data closely with slight deviations, in contrast to the Yoon-Nelson and Bohart-Adams models. The observed model validity in Table 18 is Thomas > Bohart-Adams > Yoon-Nelson. Thomas model was the best performing model for this approach, followed by the Bohart-Adams model.

Table 18. Summary of water treatment performance parameters in comparing Yoon-Nelson, Thomas and Bohart-Adams models for the Fe-seashell system during application of Global means of the empirical constants.

Parameters	Experimental		Yoon – Nelson		Thomas		B – A	
	V_b [L]	q_b [mg]	V_b [L]	q_b [mg]	V_b [L]	q_b [mg]	V_b [L]	q_b [mg]
Mass [g]								
30	*0.745*	301.1	2.78	1123	*0.755*	305.1	*0.48*	194.0
60	*1.84*	743.6	2.78	1123	*2.24*	905.3	*2.65*	1071
90	*3.125*	1263	2.78	1123	*3.725*	1505	*4.815*	1946
Flow rate [mL/min]								
2.5	*2.2*	889.1	1.39	561.7	*2.605*	1053	*3.49*	1410
5	*1.825*	737.5	2.78	1123	*2.97*	1200	*2.65*	1071
7.5	*1.425*	575.9	4.17	1685	*1.875*	757.7	*1.808*	730.5
Concentration [mg/L]								
233.22	*4.85*	1075	2.78	615.9	*4.085*	905.1	*4.83*	1070
308.4	*3.025*	886.3	2.78	814.5	*3.09*	905.3	*3.65*	1069
425.4	*1.85*	747.6	2.78	1123	*2.24*	905.3	*2.65*	1071

4.3.2 Pb-charcoal system

The generalization approach was also applied on the Pb-charcoal system and similar trends were observed for all models under consideration. The generalized model parameters for all considered models are tabulated in Table 19. The approach evaluated model performance under generalized conditions and statistical analysis results were tabulated in Table 20.

Table 19. Global means of the empirical constants values applied during the prediction of breakthrough curves for Pb-charcoal system.

Thomas		Yoon-Nelson		B-A	
K_{TH,ave_2} x 10^{-5} [L/min.mg]	q_{o,ave_2} [mg/g]	K_{YN,ave_2} [min^{-1}]	τ_{ave_2} [min]	K_{B,ave_2} x 10^{-5} [L/min.mg]	N_{o,ave_2} [mg/L]
±80.478	±58.405	±0.0147	±493.024	±48.133	±9647.679

The predictions of the Yoon-Nelson model, refitting is warranted in order to obtain a new set of model parameters when operating conditions change. The Thomas model predictions showed sensitivity when generalized mean values were applied, furthermore, refitting when operating conditions change can be mitigated since the effects of varying operating conditions was evident when the Thomas model parameters were generalized.

Materials Research Forum LLC
https://doi.org/10.21741/9781644901397-5

Table 20. Summary of performance evaluation metrics in comparing Yoon-Nelson, Thomas and Bohart-Adams models for Pb-charcoal system during application of the Global means of the empirical constants.

Parameters	Yoon – Nelson			Thomas			B – A		
	MAE	RMSE	R^2	MAE	RMSE	R^2	MAE	RMSE	R^2
Mass [g]									
1.12	0.158	0.293	*0.550*	0.006	0.011	*0.999*	53.90	100.3	*0.116*
2.52	0.030	0.075	*0.971*	0.030	0.074	*0.972*	0.325	0.811	*0.282*
3.501	0.063	0.147	*0.887*	0.035	0.083	*0.956*	0.012	0.029	*0.526*
Flow rate [mL/min]									
10	0.057	0.137	*0.902*	0.058	0.141	*0.846*	0.005	0.013	*0.508*
15	0.030	0.076	*0.970*	0.030	0.075	*0.972*	0.325	0.810	*0.282*
20	0.114	0.262	*0.643*	0.061	0.141	*0.893*	2.767	6.384	*0.142*
Concentration [mg/L]									
10	0.030	0.072	*0.973*	0.033	0.080	*0.611*	0.002	0.006	*0.694*
20	0.030	0.077	*0.970*	0.030	0.076	*0.971*	0.323	0.805	*0.280*
30	0.085	0.204	*0.785*	0.009	0.021	*0.998*	34.28	82.62	*0.113*

Table 21. Summary of water treatment performance parameters in comparing Yoon-Nelson, Thomas and Bohart-Adams models for the Pb-charcoal system during application of Global means of the empirical constants.

Parameters	Experimental		Yoon – Nelson		Thomas		B – A	
	V_b [L]	q_b [mg]	V_b [L]	q_b [mg]	V_b [L]	q_b [mg]	V_b [L]	q_b [mg]
Mass [g]								
1.12	*0.975*	18.53	7.395	140.5	*0.54*	10.26	*0.225*	4.275
2.52	*4.62*	87.78	7.395	140.5	*4.62*	87.78	*7.56*	143.6
3.501	*6.75*	128.3	7.395	140.5	*7.485*	142.2	*12.45*	236.6
Flow rate [mL/min]								
10	*4.5*	85.5	4.93	93.67	*5.53*	105.1	*5.04*	95.76
15	*4.62*	87.8	7.395	140.5	*4.62*	87.78	*13.67*	259.6
20	*3*	57	9.86	187.3	*3.72*	70.68	*6*	114
Concentration [mg/L]								
10	*6.375*	60.56	7.395	70.25	*9.24*	87.78	*15.12*	143.6
20	*4,5*	85,5	7.395	140.5	*4.62*	87,78	*7.56*	143.6
30	*2,85*	81,2	7.395	210.8	*3.09*	88,1	*5.04*	143,6

The observed error metrics are significantly higher for the generalized Bohart-Adams model when compared to those of the generalized Thomas model as seen in Table 20. Overall, the generalized Thomas model outperformed the generalized Bohart-Adams and Yoon-Nelson models in predicting the breakthrough curves as shown by higher R^2 values lower values of MAE and RMSE in Table 20. The observed model validity in Table 20 under generalized conditions for the Pb-charcoal is Thomas > Bohart-Adams > Yoon-Nelson.

Table 21 show experimental and predicted performance parameters after the application of *Global means* of model parameters for the Pb-charcoal system. The results observed show that the Thomas model tracks the experimental data very well, in contrast to the Yoon-Nelson and Bohart-Adams models. The observed model validity in Table 21 is Thomas > Bohart-Adams > Yoon-Nelson.

4.3.3 Pb-Agirus bisporus system

The generalization approach was lastly applied to the Pb-*Agirus bisporus* system and similar trends were observed for all models under consideration. The generalized model parameters for are tabulated in Table 22. The statistical analysis and water treatment performance parameters results are tabulated in Tables 23 and 24, respectively.

Table 22. Global means of the empirical values applied during the prediction of breakthrough curves for Pb-Agirus bisporus system.

Thomas		Yoon-Nelson		B-A	
K_{TH,ave_2} x 10^{-5} [L/min.mg]	q_{o,ave_2} [mg/g]	K_{YN,ave_2} [min^{-1}]	τ_{ave_2} [min]	K_{B,ave_2} x 10^{-5} [L/min.mg]	N_{o,ave_2} [mg/L]
±62,4	±63,4	±0,030	±431	±55,9	±6810,24

The concentration-time profiles demonstrating the efficacy of the considered generalized dynamic adsorption models in predicting experimental data at different sorbent masses, flow rates and initial sorbate concentrations are presented as follows.

The generalized Thomas model proved to be superior at predicting experimental data, in contrast to the generalized Bohart-Adams and Yoon-Nelson models as displayed by the mean absolute error (MAE) and root mean squared (RMSE) presented in Table 23. The generalized Bohart-Adams model was seen failing to mimic the experimental data in most cases from all cases studied, as a result, the error metrics were significantly higher when compared to those of the generalized Thomas model as seen in Table 23. The observed model validity in Table 23 under generalized conditions for the Pb-*Agirus bisporus* system is Thomas > Bohart-Adams > Yoon-Nelson. Table 24 show experimental and predicted performance parameters after the application of generalized mean values of model parameters for the Pb-*Agirus bisporus* system. The results observed show that the generalized Thomas model tracks the experimental data very well, in contrast to the

Materials Research Forum LLC
https://doi.org/10.21741/9781644901397-5

generalized Yoon-Nelson and Bohart-Adams models. The observed model validity in Table 24 is Thomas > Bohart-Adams > Yoon-Nelson.

Table 23. Summary of performance evaluation metrics in comparing Yoon-Nelson, Thomas and Bohart-Adams models for Pb-Agirus bisporus system during application of Global means of the empirical constants.

Parameters	Yoon – Nelson			Thomas			B – A		
	MAE	RMSE	R^2	MAE	RMSE	R^2	MAE	RMSE	R^2
Mass [g]									
0.4	0.006	0.010	*0.410*	0.048	0.080	*0.962*	15.93	26.60	*0.439*
0.8	0.089	0.171	*0.818*	0.012	0.023	*0.997*	143.5	277.4	*0.265*
1.2	0.049	0.101	*0.953*	0.034	0.070	*0.974*	16949	34889	*0.278*
Flow rate [mL/min]									
1	0.170	0.410	*0.168*	0.041	0.100	*0.940*	1680	4066	*0.292*
3	0.084	0.164	*0.839*	0.023	0.045	*0.990*	190.6	370.1	*0.255*
5	0.024	0.041	*0.319*	0.022	0.037	*0.992*	26.13	44.21	*0.325*
Concentration [mg/L]									
20	0.172	0.374	*0.375*	0.021	0.045	*0.978*	0.762	1.662	*0.625*
50	0.090	0.169	*0.831*	0.020	0.038	*0.993*	166.3	312.6	*0.274*
100	0.006	0.011	*0.427*	0.090	0.145	*0.902*	2404	3871	*0.253*

Table 24. Summary of water treatment performance parameters in comparing Yoon-Nelson, Thomas and Bohart-Adams models for the Pb-Agirus bisporus system during application of Global means of the empirical constants.

Parameters	Experimental		Yoon – Nelson		Thomas		B – A	
	V_b [L]	q_b [mg]	V_b [L]	q_b [mg]	V_b [L]	q_b [mg]	V_b [L]	q_b [mg]
Mass [g]								
0.4	*0,348*	16,5	1,008	48	*0,225*	10,7	*0,162*	7,7
0.8	*0,735*	34,9	1,008	48	*0,732*	34,8	*0,642*	30
1.2	*1,041*	49	1,008	48	*2,139*	102	*0,546*	26
Flow rate [mL/min]								
1	*0,94*	44,7	0,336	16,0	*0,92*	44	*0,856*	41
3	*0,72*	34,2	1,008	48	*0,732*	35	*0,642*	30
5	*0,7*	33,3	1,68	80	*0,545*	25,9	*0,430*	20,4
Concentration [mg/L]								
20	*1,8*	34	1,008	19,2	*1,83*	34,8	*1,605*	30
50	*0,735*	34,9	1,008	48	*0,732*	34,8	*0,642*	30
100	*0,39*	37,1	1,008	95,8	*0,366*	34,8	*0,321*	30

Conclusions and recommendations

Numerous mathematical models have been developed to predict dynamic adsorption performance but evaluation and efficiency of these models for large-scale operations have not been studied. In the current study, three models were evaluated to analyse the behaviour of sorbent-sorbate systems under different conditions. The considered models were namely, Yoon-Nelson, Thomas and Bohart-Adams model. The shape of the breakthrough curves contained data about the mass transfer properties of each adsorptive-adsorbent system such as breakthrough point, saturation point, maximum loading of adsorptive material, serviceable sorption capacity, among other properties. These properties were evaluated by applying modified dynamic adsorption models and fitting to experimental data by simulations. The study was aimed at providing unambiguous approaches in selecting the best performing model for dynamic adsorption systems. Predictive and generalization performances of the models were evaluated using the statistical criteria of MAE, RMSE and R^2. Thomas and Yoon-Nelson models predicted experimental data satisfactory prior to application of mean values for all adsorption systems. Based on the findings, the present study revealed that statistical measures are of significance when evaluating model performance in dynamic adsorption systems. The observed model validity prior to application of mean values was Thomas > Yoon-Nelson > Bohart-Adams. The regression results for Yoon-Nelson and Thomas models showed close relation between experimental and theoretical breakthrough curves as the regression values were close to 1, in contrast with the Bohart-Adams model that yielded low regression values. All predictions for the Yoon-Nelson model were insensitive to the application of mean values, i.e., all predicted breakthrough curves were similar despite changing operating conditions. The trends observed for the Yoon-Nelson model when mean values were applied does not support the novelty of the chosen approach. The Thomas model on the other hand was capable of predicting successfully when empirical constants were fixed for each variable range and for the entire experimental domain. The sensitivity observed for the Thomas model can be attributed to the number of operating parameters incorporated in the model in contrast to the Yoon-Nelson model. All predicted breakthrough curves deviated from the ideal s-shape form for the Bohart-Adams model when mean values were applied, which contributed to high values of error metrics (MAE and RMSE) and low R^2 values. The Local mean values (LMV) approach proved to be very pertinent for dynamic models under evaluation, particularly for the Thomas model. The Global mean approach (GMV) revealed that Yoon-Nelson model does not depend on operating conditions when empirical constants are generalized. Refitting when operating conditions change can be mitigated for the generalized Thomas model since the effects of varying operating conditions was evident when model parameters were generalized. Model validity assessed by MAE and

Materials Research Foundations **102** (2021) 128-167 | https://doi.org/10.21741/9781644901397-5

RMSE and R^2 was observed to be Thomas > Bohart-Adams > Yoon-Nelson when Local mean values (LMV) and Global mean values (GMV) were applied. Thomas model was the best performing model in all three approaches considered. The kinetic constant, K_{TH}, for the Thomas model is the most important parameter and is responsible for steepening or smoothing of the breakthrough curve. The application of mean values revealed that even slight deviations in model parameters can result in a significant shift in the concentration-time profile. Hence, in order for the application of mean values to be more accurate, most especially the local means, operating variables must not be dispersed significantly, for instance, investigating the sorbent mass at 2, 3 and 4 g is better than 2, 4 and 6 g when considering applying the local mean values of model parameters. The Bohart-Adams model was developed to predict the initial part of the breakthrough curve, however, the model failed to outperform the Thomas model in predicting the breakthrough point performance parameters. The under performance of the Bohart-Adams model, revealed that the number of operating variables incorporated in the model cannot be used to classify the overall efficacy of the model. A trade off, however, exists when considering practical applications of the model i.e. a model with more process variables is preferred over the one with no operating variables. The nature of the Thomas model predictions after application of mean values were reflective of a horizontal translation i.e. $t \pm \alpha$, where α is a shifting time constant. It is worth emphasizing that, though a significant shape of a breakthrough curve is s-shaped, the most vital section of the curve is before breakthrough point as it determines the amount of water treated, thus correcting the breakthrough section is more significant for the Thomas model. The results of the neural network model for all operating parameters under study, were closely correlated, with the actual results as indicated by the R^2 values of above 0.98 and low values of MAE and RMSE. Overall, the neural network model outperformed the generalized Thomas model in predicting the breakthrough curves for all considered adsorption systems within the range of the experimental data. Artificial neural networks (ANN) are very powerful in predicting complex non-linear problems as indicated by close correlation between actual and predicted breakthrough curves. However, since the ANN methodology uses standardized equations which do not have any methodological background, the calculated model can only be applied within experimental range and cannot be used for extrapolation. In addition, ANN requires large number of experimental data prior to training in order to obtain a highly reliable model, but this limitation escalates further as there is no method available to determine minimum number of experiments required for ANN training.

Acknowledgements

The authors acknowledge the financial support for this work by the National Research Foundation (NRF) of South Africa and the Institute of Systems Science of the Durban University of Technology, South Africa.

References

[1] Kefeni, K. K., Msagati, T. A. and Mamba, B. B. 2017. Acid mine drainage: prevention, treatment options, and resource recovery: a review. Journal of Cleaner Production, 151: 475-493. https://doi.org/10.1016/j.jclepro.2017.03.082

[2] Nleya, Y., Simate, G. S. and Ndlovu, S. 2016. Sustainability assessment of the recovery and utilisation of acid from acid mine drainage. Journal of Cleaner Production, 113: 17-27. https://doi.org/10.1016/j.jclepro.2015.11.005

[3] Ali, I. 2014. Water treatment by adsorption columns: evaluation at ground level. Separation & Purification Reviews, 43 (3): 175-205. https://doi.org/10.1080/15422119.2012.748671

[4] Podstawczyk, D., Witek-Krowiak, A., Dawiec, A. and Bhatnagar, A. 2015. Biosorption of copper (II) ions by flax meal: Empirical modeling and process optimization by response surface methodology (RSM) and artificial neural network (ANN) simulation. Ecological Engineering, 83: 364-379. https://doi.org/10.1016/j.ecoleng.2015.07.004

[5] Xu, Z., Cai, J.-g. and Pan, B.-c. 2013. Mathematically modeling fixed-bed adsorption in aqueous systems. Journal of Zhejiang University SCIENCE A, 14 (3): 155-176. https://doi.org/10.1631/jzus.A1300029

[6] Calero, M., Hernáinz, F., Blázquez, G., Tenorio, G. and Martín-Lara, M. 2009. Study of Cr (III) biosorption in a fixed-bed column. Journal of hazardous materials, 171 (1-3): 886-893. https://doi.org/10.1016/j.jhazmat.2009.06.082

[7] Turan, N. G., Mesci, B. and Ozgonenel, O. 2011a. Artificial neural network (ANN) approach for modeling Zn (II) adsorption from leachate using a new biosorbent. Chemical Engineering Journal, 173 (1): 98-105. https://doi.org/10.1016/j.cej.2011.07.042

[8] Oguz, E. and Ersoy, M. 2010b. Removal of Cu 2+ from aqueous solution by adsorption in a fixed bed column and Neural Network Modelling. Chemical Engineering Journal, 164 (1): 56-62. https://doi.org/10.1016/j.cej.2010.08.016

[9] Kumar, D., Pandey, L. K. and Gaur, J. 2016. Metal sorption by algal biomass: From batch to continuous system. Algal research, 18: 95-109. https://doi.org/10.1016/j.algal.2016.05.026

[10] Geankoplis, C. J. 2003. Transport processes and separation process principles:(includes unit operations). Prentice Hall Professional Technical Reference.

[11] Bohart, G. and Adams, E. 1920. Some aspects of the behavior of charcoal with respect to chlorine. Journal of the American Chemical Society, 42 (3): 523-544. https://doi.org/10.1021/ja01448a018

[12] Chu, K. H. 2010. Fixed bed sorption: setting the record straight on the Bohart–Adams and Thomas models. Journal of hazardous materials, 177 (1-3): 1006 1012. https://doi.org/10.1016/j.jhazmat.2010.01.019

[13] Hutchins, R. 1973. New method simplifies design of activated-carbon systems. Chemical Engineering, 80 (19): 133-138.

[14] Ahmad, A. and Hameed, B. 2010. Fixed-bed adsorption of reactive azo dye onto granular activated carbon prepared from waste. Journal of hazardous materials, 175 (1): 298-303. https://doi.org/10.1016/j.jhazmat.2009.10.003

[15] Han, R., Wang, Y., Zhao, X., Wang, Y., Xie, F., Cheng, J. and Tang, M. 2009. Adsorption of methylene blue by phoenix tree leaf powder in a fixed-bed column: experiments and prediction of breakthrough curves. Desalination, 245 (1-3): 284-297. https://doi.org/10.1016/j.desal.2008.07.013

[16] Thomas, H. C. 1944. Heterogeneous ion exchange in a flowing system. Journal of the American Chemical Society, 66 (10): 1664-1666. https://doi.org/10.1021/ja01238a017

[17] Yoon, Y. H. and Nelson, J. H. 1984. Application of gas adsorption kinetics I. A theoretical model for respirator cartridge service life. The American Industrial Hygiene Association Journal, 45 (8): 509-516. https://doi.org/10.1080/15298668491400197

[18] Masukume, M., Onyango, M. S. and Maree, J. P. 2014. Sea shell derived adsorbent and its potential for treating acid mine drainage. International Journal of Mineral Processing, 133: 52-59. https://doi.org/10.1016/j.minpro.2014.09.005

[19] Biswas, S. and Mishra, U. 2015. Continuous fixed-bed column study and adsorption modeling: Removal of lead ion from aqueous solution by charcoal originated from chemical carbonization of rubber wood sawdust. Journal of Chemistry, 2015. https://doi.org/10.1155/2015/907379

[20] Long, R. A. 2010. Continuous fixed-bed column study and adsorption modeling: Removal of lead ion from aqueous solution by Agirus Bisporus. Journal of Biomedical Materials Research Part A, 103 (3): 949-958.

Materials Research Foundations **102** (2021) 168-181
https://doi.org/10.21741/9781644901397-6

Chapter 6

Hydrolytic Acidification-Two Stage EGSB-A/O Combined Process for Pesticide Wastewater Treatment

Yongjun Sun*, Shengbao Zhou, Kinjal J. Shah*, Xuefeng Xiao
College of Urban Construction, Nanjing Tech University, Nanjing, 211816, China
Yongjun Sun: sunyongjun@njtech.edu.cn, Kinjal J. Shah: kjshah@njtech.edu.cn

Abstract

The wastewater produced by a chemical enterprise in Ningxia has the characteristics of high COD, high salt content, and high ammoniacal nitrogen. The biodegradability of the wastewater is poor due to presence of higher concentration of pollutant. The scale of first stage of treatment was $400m^3/d$. The wastewater of different characteristic was collected and treated separately. The combined process of "micro electrolysis + Fenton oxidation + coagulation precipitation + evaporation crystallization" was used to pretreat the wastewater containing high salt and high COD. The main process was "hydrolytic acidification + two-stage Expanded Granular Sludge Bed (EGSB) + two-stage advance Fenton oxidation (A/O)". The final concentrations of effluent COD, NH_3-N and TDS are 376 mg/L, 34.74 mg/L, and 442 mg/L, and the removal rates are 99.9%, 81.3%, and 99.9%, respectively. The environmental, engineering and economical (3E) practices showed that the combined process has stable operation, strong impact resistance, and high reduction of COD in the effluent.

Keywords

Pesticide Wastewater, Pretreatment, Two-Staged EGSB, Advance Oxidation, Fenton Oxidation

Contents

1. Introduction

Pesticides are important chemical substances used in the prevention and control of plant diseases and insect pests in agricultural production. [1]. In recent years, through the advancement of science and technology and the rapid development of the green agricultural practices, China has started their move towards major producer of pesticides [2]. Therefore, the pesticide industry has become one of the important industries in China's industrial system [2]. According to the available statistics, there are currently more than 2,500 pesticide manufacturing units/industries in China. While comparing those with developed countries, there is still a big gap in the overall level of China's pesticide industry, mainly manifested in more outdated and less ultra-efficient varieties, irrational product structure, environmental hazards, and toxicity [3]. Among them, many manufacturers are reluctant to invest much in research and development, and thus they are practicing with old or expired manufacturing methods [4]. On the other hand, some products have been eliminated in the international production process or even banned due to implementing green chemical and hazardous free production practices. Now these low-end products are not only harmful to the environment and toxic but also cause fierce market competition, leading to a bleak market for new products, which leads to a vicious cycle [5].

While considering the environmental impacts, the total amount of wastewater discharged from the pesticide production process is about 300 million tons per year [6]. These wastewaters have high pollutant concentration, deep color, high salt content, complex ingredients, and high toxicity, and poor biodegradability. If it is directly discharged without effective treatment, it will not only pollute surface water but also pollute the ground water. Moreover, eutrophication effects also will be found after discharging polluted water to other water system, which cause great harm to various animals and plants in the water body [7]. It is very difficult to set up protocols for chemical waste in water system because the raw materials and production processes are different for variety of pesticides. This will lead to generates complex wastewater composition. In ordered to identify it and categorized it, few generalized features have been listed below:

a) The concentration of organic pollutants is quite high: far exceeding the national takeover standards allowed. The value of chemical oxygen demand can reach tens of thousands of mg/L, and the concentration of organic matter that is difficult to degrade is beyond the range that the sewage plant can take over [8].

b) The biological toxicity is very high: the wastewater contains not only some pesticides, pesticide intermediates, and difficult-to-degrade substances, but also high concentrations of toxic and harmful substances such as phenol, arsenic, and mercury [9].

c) It has a strong pungent odor: it has severe irritation and damage to the respiratory organs of the human body and has a certain impact on the surrounding ecological environment [10].

d) The composition and content of wastewater and the amount of wastewater are not easy to control: the time and type of products production are related to market demand and seasonal changes [11].

Pesticide wastewater contains a large amount of toxic and hazardous substances such as nitrobenzene, chloronitrobenzene, chlorobenzene, etc. Thus, the complex types of pesticide have complex composition of wastewater and a single treatment method to achieve pure water is very expensive [12]. Meanwhile, the high concentration of refractory organic substances in it increases the treatment complexity, increase the treatment cost, and greatly increase the risk of the sewage treatment plant exceeding the standard [14]. If the concentration of polluted water is not controlled upon time, it will damage to the soil and which leads to enter into food chain system to affect human health. Meanwhile, the untreated pesticide water directly enters the sewage treatment plant, and the toxic and harmful substances have a destructive effect on the activated sludge in the biological reaction tank, reducing its activity and processing capacity which causes a large number of deaths of microorganisms in activated sludge. [13].

There are many processes for treating pesticide wastewater or high-salt and high-concentration organic wastewater, but it is very difficult to achieve the desired treatment effect by using a certain physical and chemical method or biochemical method alone.

Therefore, according to the characteristics of wastewater and the advantages and disadvantages of various treatment methods, many studies have been carried out with a variety of physicochemical and biochemical combined processes to treat wastewater. Generally, the beginning of the processes has been carried out by physical and chemical methods where suspended matter and colloid in the wastewater are removed. If the salt content of the wastewater is too high, it needs to be desalted. Followed by advance oxidation technology to reduce the toxicity of wastewater. Additionally, the biodegradability of wastewater was carried out by biochemical treatment processes to meet the influent requirements. Finally, through biodegradation, the organic matter in the wastewater is degraded to ensure that the final effluent quality meets the discharge requirements. If the standard cannot be reached after biochemical treatment, the wastewater needs to be further treated. In such case, the combined process has been widely used in the field of industrial wastewater treatment, but for China's pesticide wastewater, due to the complexity and variety of products and processes, the types of pollutants in wastewater are often very different. Different combination of processes should be used according to the characteristics of wastewater to achieve the best treatment effect.

This research was carried out on wastewater samples of fine chemical manufacturing company, Ningxia, producer of pesticides such as propachlor, metalaxyl, difenoconazole, and pesticide intermediates such as dimethalone and toluene. The wastewater is collected as per different segment of the plants namely, liquor wastewater, toluene wastewater, alkaline wastewater, neutralization wastewater, and domestic sewage in the plant. The compositions of those wastewater plants are complex, the concentration of organic matter is high, the content of COD and NH_3-N is high, and the salt content is high (See Table 1). Thus, it is difficult to achieve the expected results with only applying biochemical treatment methods. But, after applying pretreatment process, biochemical treatment, and advanced treatment process of the wastewater jointly, the effluent meets the Class B standard of "Water Quality Standards for Wastewater Discharge into Urban Sewers" [GBT 31962-2015] and meets the requirements for access to the sewage pipe network of the park. The specific water quality of the inlet and outlet is shown in Table 1.

Materials Research Foundations **102** (2021) 168-181 https://doi.org/10.21741/9781644901397-6

Table. 1 Design influent and effluent water quality.

Wastewater name	Water volume [m^3/d]	COD [mg/L]	NH_3-N[mg/L]	TN [mg/L]	TP [mg/L]	pH	Salt content [mg/L]
Heptenone acid water	44.2	477500	129.03	212.89	0.713	7.32	89600
Synthetic acid water	7.8	243500	1146.75	1651.32	10.746	1.11	9550
Hypochlorous toluene wash	14.4	487750	127.63	243.77	0.19	10.29	70360
Liquid alkaline washing with toluene	39.8	188000	79.03	126.37	0.19	11.55	62320
Acetochloric acid water	63.8	2	28.38	54.77	0.143	1.28	52975
Clethodim one wash	32.4	2000	6.81	10.87	0.095	5.25	149
Decarboxylated water	77.2	41600	738.06	1320.18	1.664	0.75	941
Decarboxylation once washed	24.5	88500	32.44	57.61	0.285	1.77	2380
Synthetic one wash	26.2	3740	258.75	387.6	0.333	2.72	2200
Synthetic secondary washing	25.4	7890	254.85	389.21	0.095	4.78	248
Wash toluene once	21.5	22280	32.75	59.28	0.095	6.67	647
Washing toluene twice	12.6	2300	39	74.5	0.024	3.78	194
Heptenone wash once	9.6	74200	171.84	307.25	0.38	4.91	3730
Design effluent quality	400	500	45	70	8	6~9	—

2. Treatment process selection

2.1 Pretreatment process selection

It can be seen from Table 1 that multiple streams of wastewater belong to high-salt and high-COD wastewater category. As discussed before, multi component wastewater require specific pretreatment. In case of high-salt and high-COD category, Fe/C micro-electrolysis coupled Fenton redox-oxidation of wastewater can be used which can degrade macromolecular organics to make small and simpler molecules for improving the wastewater biodegradability. However, it has a limitation of controlling the amount of Fe^{2+} and Fe^{3+}, the large amount of Fe^{2+} and Fe^{3+} remaining in the wastewater is not conducive to the subsequent biochemical treatment. Thus, this technique needed follow-up techniques. In that case, coagulation and precipitation techniques can be implemented where Fe^{2+} and Fe^{3+} of the solution converted to form of $Fe[OH]_2$ and $Fe[OH]_3$. However, the salinity in the wastewater is high, which is not good for the microorganisms in the subsequent biochemical unit. Thus, the evaporative salt precipitation is used to treat high-salt wastewater.

2.2 Biochemical treatment process selection

The wastewater after pretreatment still has a high concentration of organic matter. Therefore, the biochemical treatment adopts the process of "hydrolytic acidification two-stage Expanded Granular Sludge Bed (EGSB) +A/O". Studies have shown that hydrolytic acidification can convert difficult-to-biodegradable macromolecular substances into easily biodegradable substances, further improving the wastewater biodegradability. The EGSB anaerobic treatment process is an anaerobic biological treatment process specially designed for high-concentration organic industrial wastewater. It has a high volume load and impact resistance, and can effectively resist fluctuations in water quality, and it is simple to operate and very practical treatment process. In addition, the combination of hydrolytic acidification and EGSB provides short-term denitrification conditions for anaerobic ammonia oxidation, which can effectively reduce the ammoniacal nitrogen load, reduce the need for external carbon sources due to deamination, and achieve the system's requirements for the ammoniacal nitrogen index. Due to the low TP index in the wastewater and the high concentration of ammoniacal nitrogen and COD, after the load is greatly reduced by EGSB, the denitrifying bacteria in the anoxic tank will be used as the carbon source for the COD reduction in the wastewater to bring the aerobic tank back into the mixture. NO_3-N and NO_2-N are converted into N_2 by denitrification, and the concentration of BOD_5 and NO_3-N are reduced. Finally, the biodegradable COD and ammoniacal nitrogen are completely degraded.

2.3 Advanced treatment process selection

The Fenton oxidation process is selected as the advanced treatment process, which can not only increase the impact load caused by the change in water volume but also achieve efficient precipitation in the sedimentation tank. In summary, the process flow of the wastewater treatment scheme is shown in Figure 1.

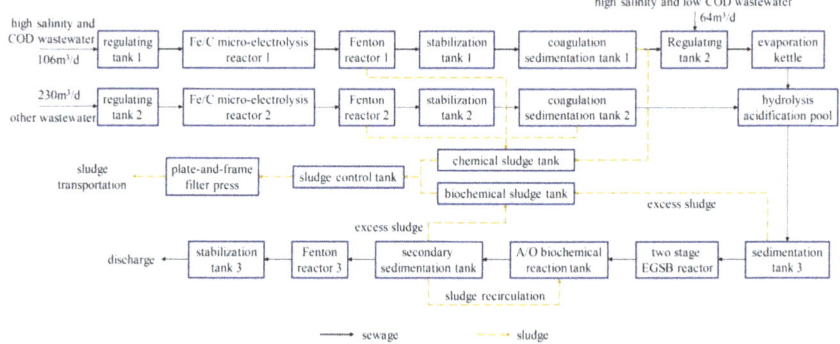

Figure. 1 Flow chart of the Wastewater Treatment Process.

2.4 Wastewater treatment process

The wastewater treatment process flow is shown in Fig. 1. Wherein, 106 m^3/d high-salt and high-COD wastewater and 230 m^3/d other wastewater are pretreated in two ways. As shown in Fig. 1, the wastewater first enters the regulating tanks 1 and 2 to regulate the water quality and quantity where the pH is adjusted to 3-4. Followed by, the wastewater flows into the Fe/C micro electrolysis reactors 1 and 2 to reduce and degrade the organic pollutants in the wastewater. Subsequently, the wastewater flows into the Fenton reactors 1 and 2 by gravity. Here, the hydroxyl radicals oxidize and degrade pollutants to improve the biodegradability of organic pollutants. The active groups of wastewater are decomposed completely. After Fenton process, effluent is passed to stabilization tank 1 and 2, where pH is adjusted to 8-9 by the pH-adjusting reactor. Now the effluent flows into the coagulation sedimentation tanks 1 and 2 to complete the flocculation reaction where separation is taking place into sludge and water. The remaining sludge is pumped into the chemical sludge tank for further treatment. The effluent having 64 m^3/d of high-salt and low-COD flow into the regulating tank 3 for the evaporation kettle. Here, the concentrated liquid enters the evaporator and evaporates most of the wastewater. Finally, it is cooled and crystallized there. The wastewater from the coagulation sedimentation tank 2 enters the hydrolysis and acidification tank where degradation of macromolecular organic matter taking place into small-molecule organic matter to improve biodegradability. The effluent from the hydrolysis and acidification tank flows into the precipitation tank 3 to prevent the growth of micro-organisms in the subsequent biochemical treatment unit. The effluent from the precipitation tank 3 enters the two-stage EGSB reactor. Subsequently, the wastewater enters the biochemical reaction tank, degrades the BOD of the wastewater and

Materials Research Foundations **102** (2021) 168-181
https://doi.org/10.21741/9781644901397-6

then the wastewater flows into the secondary precipitation tank to realize the separation of sludge and water. Part of the sludge in the precipitation tank 3 and the secondary sedimentation tank is returned to the hydrolysis acidification tank and the anoxic/aerobic biochemical reaction tank, and the excess sludge is sent to the biochemical sludge tank for treatment. The treated wastewater from the secondary precipitation tank enters the pH-adjusting reaction kettle by gravity-flow, and the pH is adjusted to 3-4 by dosing the concentrated H_2SO_4, and the effluent is pumped to the Fenton reactor 3 for advanced treatment of wastewater. In the final stage, the sludge in the biochemical sludge tank and the chemical sludge tank is conditioned by the sludge adjustment tank, and then filtered by a plate and frame filter press to produce sludge cake which is transported out, and the filtered water returns to the adjustment tank.

3 Main processing unit and design parameters

3.1 Pretreatment system

[1] Adjusting tank 1. Two seats, design flow Q_1 = 106 m^3/d, size Φ3.0 m × 3.0m [including 0.5 m super high], effective volume 35.33 m^3, HRT = 8 h. Adopt carbon steel lining plastic anti-corrosion structure. Supporting equipment: 2 pH online instruments, 2 flow meters; 2 submersible mixers, power N = 0.35 kW.

[2] Fe/C micro-electrolysis reactor 1. Two seats, size Φ3.0 m × 2.0 m [including 0.5 m super high], effective volume 17.67 m^3, HRT = 4 h. Adopt carbon steel lining plastic anti-corrosion structure. Supporting equipment: 2 sets of aeration device iron carbon filler 9 m^3.

[3] Fenton reactor 1. Two seats, size Φ3.0 m × 2.0 m [including 0.5 m super high], effective volume 17.67 m^3, HRT = 4 h. Adopt carbon steel lining plastic anti-corrosion structure. Supporting equipment: 2 pulp mixers, model JBJ-800 [D = 800 mm, N = 1.0 kW].

[4] 1.2 stable pools, size Φ3.0 m × 2.0 m [including 0.5 m super high], effective volume 17.67 m^3, HRT = 4 h. Adopt carbon steel lining plastic anti-corrosion structure. Supporting equipment: 2 pulp mixers, model JBJ-800 [D = 800 mm, N = 1.0 kW].

[5] Coagulation tank 1. One, with a size of 2.0 m × 1.2 m × 1.7 m [including an ultra-high 0.5 m], an effective volume of 2.95 m^3, and HRT = 40 min. Adopt steel concrete anti-corrosion semi-underground structure. Supporting equipment: one set of gas stirring device, one tube static mixer, model SK-25 / 50 [DN = 50, Q = 3.5 ~ 7.0 m^3/h].

[6] Settling tank 1. One, with a size of 2.8 m × 2.0 m × 2.8 m [including an ultra-high 0.5 m], an effective volume of 13.25 m^3, HRT = 3 h, and a surface load q=0.8 $m^3/[m^2h]$. Adopt steel concrete anti-corrosion semi-underground structure. Supporting equipment: 1 set of

Materials Research Foundations **102** (2021) 168-181
https://doi.org/10.21741/9781644901397-6

scraper and 2 sets of sludge pumps (Q=3.0 m^3/h, H=10 m, N=0.55 kW), One used, one spare.

[7] Adjusting tank 2. Two, design flow Q2 = 170 m^3/d, size Φ3.0 m × 4.3 m [including 0.3 m super high], effective volume 56.67 m^3, HRT = 8 h. Adopt carbon steel lining plastic anti-corrosion structure. Supporting equipment: 2 pH online instruments, 2 flow meters; 2 submersible mixers, power N = 0.35 kW.

[8] Evaporation kettle. One, design flow Q = 7.0 m^3/h. supporting equipment: 1 set of evaporation kettle equipment.

[9] Regulating tank 3. Two seats, design flow Q_3 = 230 m^3 / d, size Φ3.0 m × 5.7 m [including 0.3 m super high], effective volume 76.67 m^3, HRT = 8 h. Adopt carbon steel lining plastic anti-corrosion structure. Supporting equipment: 2 pH online instruments, 2 flow meters; 2 submersible mixers, power N = 0.35 kW.

[10] Fe / C micro-electrolysis reactor 2. Two seats, size Φ3.0 m × 3.2 m [including 0.5 m super high], effective volume 38.33 m^3, HRT = 4 h. Adopt carbon steel lining plastic anti-corrosion structure. Supporting equipment: 2 sets of aeration devices iron carbon filler 19 m^3.

[11] Fenton reactor 2. Two seats, size Φ3.0 m × 3.2 m [including 0.5 m super high], effective volume 38.33 m^3, HRT = 4 h. Adopt carbon steel lining plastic anti-corrosion structure. Supporting equipment: 2 pulp mixers, model JBJ-800 [D = 800 mm, N = 1.0 kW].

[12] 2 stable tanks, size Φ3.0 m × 3.2 m [including 0.5 m super high], effective volume 38.33 m^3, HRT = 4 h. Adopt carbon steel lining plastic anti-corrosion structure. Supporting equipment: 2 pulp mixers, model JBJ-800 [D = 800 mm, N = 1.0 kW].

[13] Coagulation tank 2. One, with a size of 2.5 m × 1.5 m × 2.2 m [including an ultra-high 0.5 m], an effective volume of 6.39 m^3, and HRT = 40 min. Adopt steel concrete anti-corrosion semi-underground structure. Supporting equipment: one set of gas stirring device, one tube static mixer, model SK-25 / 50 [DN = 50, Q = 3.5 ~ 7.0 m^3/h].

[14] Settling tank 2. One, with a size of 4.0 m × 3.0 m × 2.9 m [including an ultra-high 0.5 m], an effective volume of 28.75 m^3, HRT = 3 h, and a surface load q = 0.8 m^3 / [m^2 • h]. Adopt steel concrete anti-corrosion semi-underground structure. Supporting equipment: 1 set of scraper and 2 sets of sludge pumps [Q = 3.0 m^3/h, H = 10 m, N = 0.55 kW], one for each use.

3.2 Biochemical treatment system

[1] Hydrolysis acidification tank. One, with a design flow of 400 m^3/d, a size of 20.0 m × 8.0 m × 6.0 m [including an ultra-high 0.5 m], volumetric load N_V = 3.0 kg/[m^3 • d], and an effective volume of 856.4 m^3. Adopt steel concrete anti-corrosion semi-underground structure. Auxiliary equipment: 1 pH online instrument, 1 ORP online instrument; 2 push flow stirring devices, power N = 3 kw. There are 4 front tanks in the hydrolytic acidification tank, with a size of 10.0 m × 5.0 m × 4.5m [including ultra-high 0.5 m], an effective volume of 800 m^3, and HRT = 48 h. Adopt steel concrete anti-corrosion semi-underground structure. Supporting equipment: one pH online meter, one flow meter; eight submersible mixers, power N = 1.0 kw; eight lift pumps, model KQL40-100 [I] [Q = 16.3 m^3/h, H = 11.3 m, N = 1.1 kW], 4 uses 4 spares.

[2] Settling tank 3. One, with a size of 5.5 m × 3.8 m × 3.7 m [including an ultra-high 0.5 m], an effective volume of 66.67 m^3, HRT = 4 h, and a surface load q = 0.8 m^3/[m^2 • h]. Adopt steel concrete anti-corrosion semi-underground structure. Supporting equipment: 1 set of scraper and 2 sets of sludge pumps [Q = 5.0 m^3/h, H = 15 m, N = 0.75 kW], one for each use.

[3] EGSB reactor. 2 sets of 4 units, design flow Q = 400 m^3 / d, size Φ8.0 m × 24.0 m. Adopt carbon steel lining plastic anti-corrosion structure. Supporting equipment: 4 sets of pulse water replenisher, 8 sets of biogas slurry separator.

[4] Anoxic pool. One, with a size of 8.0 m × 5.5 m × 4.2 m [including an ultra-high 0.5 m], an effective volume of 164.27 m^3, and a volumetric load N_V = 1.50 kg/[m^3 • d]. Adopt steel concrete anti-corrosion semi-underground structure. Auxiliary equipment: 1 pH online instrument, 1 DO online instrument; 2 push flow stirring devices, power N = 1.50 kw.

[5] Aerobic pool. Two, size 20.0 m × 8.0 m × 5.2 m [including ultra-high 0.5 m], effective volume of 1485 m^3, volume load N_V = 0.8 kg/[m^3 • d]. Adopt steel concrete anti-corrosion semi-underground structure. Auxiliary equipment: 2 pH online instruments, 2 DO online instruments, 2 micro porous aeration systems; 4 push flow stirring devices, power N = 3.0 kw; 4 internal reflux pumps, Q = 60 m^3/h, H = 15 m, N = 5.5 kW, 2 for 2 and 2; MBBR filler 700 m^3.

[6] Second sink pool. One, with a size of Φ5.0 m × 2.7 m [including ultra-high 0.5 m], effective volume of 41.67 m^3, HRT – 2.5 h, and surface load q = 0.85 m^3/[m^2 • h]. Adopt steel concrete anti-corrosion semi-underground structure. Auxiliary equipment: 1 set of mud scraper, 2 sets of sludge pump, Q = 5.0 m^3/h, H = 15 m, N = 0.75 kW, 1 for 1 spare.

Materials Research Forum LLC
https://doi.org/10.21741/9781644901397-6

3.3 Advanced processing system

[1] Fenton reactor 3. Two seats, size Φ3.0 m × 5.2 m [including 0.5 m super high], effective volume 66.67 m^3, HRT = 4h. Adopt carbon steel lining plastic anti-corrosion structure. Supporting equipment: 2 pulp mixers, model JBJ-800 [D = 800 mm, N = 1.0 kW].

[2] Stable pool 2. Two seats, size Φ3.0 m × 5.2 m [including 0.5 m super high], effective volume 66.67 m^3, HRT = 4 h. Adopt carbon steel lining plastic anti-corrosion structure. Supporting equipment: 2 pulp mixers, model JBJ-800 [D = 800 mm, N = 1.0 kW].

4. 3E analysis of each operation units

After about 60 days of commissioning and operation, each reactor was successfully started, and the effluent water quality of some treatment structures is shown in Table 2.

Table 2 Effluent water quality of water treatment structure

Reactors	COD [mg/L]	NH4-N [mg/L]	Salt content [mg/L]	TP [mg/L]	pH
Regulating tank 1	358386	184.85	70888	1.18	8.85
Regulating tank 2	130103	176.94	42200	0.74	5.49
Evaporation kettle	65051	159.24	1688	0.65	7.24
Regulating tank 3	30080	322.98	1098	0.68	3.06
Coagulation sedimentation tank 2	17386	465.09	553	0.45	7.82
Hydrolysis acidification tank	30115	268.08	673	0.28	7.56
Two-stage EGSB reactors	3689	96.51	545	0.36	7.34
A/O Biochemical reaction tank	470	38.60	491	0.15	7.55
Fenton Reactor 3	376	34.74	442	0.03	2.78

It can be seen from Table 2 that the combined process of pretreatment + biochemical treatment + in-depth treatment finally stabilizes the COD, NH_3-N, TN, TP values. While comparing table 1 and table 2, it can be seen that there is significant difference in the values of COD of the effluent at about 400 mg /L, the ammonia nitrogen at about 35 mg / L, and the TP does not detect. By the time, they all meet the Class B standard of "Water Quality Standards for Wastewater Discharge into Cities and Towns" [GBT 31962-2015]. The effluent water quality has a small fluctuation range and the process can be operated stably. So it can be concluded that the above designed process of pretreatment + biochemical treatment + in-depth treatment works more effectively and efficiently. However, it is mandatory to consider 3E benefits, environmental, engineering and economical. Above mentioned benefits for environmental where we can found significant improvement of wastewater quality where COD was found to be 376 mg/L with 99.9% removal rate, whereas NH3-N and TDS were found to be 34.74 mg/L and 442 mg/L with 81.3 and 99.9% removal rates, respectively. Similarly, the engineering practice shows that the combined process has stable operation, strong impact resistance, and high reduction of COD in the effluent. There are 3 major engineering drawbacks which support this combined process, i.e.

[1] If we treat wastewater separately, it will not be advisable for the prospects of material safety as it has to pass from unit to unit and each pollutants have different characteristics. However, in this kind of combined process, water quality is treated in different treatment sections in a targeted manner to resist impact load and ensure the safe and stable operation of the overall process.

[2] The evaporative crystallization process is used to efficiently remove salts, which solves the effect of high-salinity wastewater on active micro-organisms in the biochemical treatment process.

[3] The Fe/C micro-electrolysis coupled Fenton oxidation method is used to retreat the wastewater. The micro-electrolysis reaction causes large molecular organics to open and break the chain while generating a large amount of Fe^{2+}, which greatly reduces the dose of Fenton reaction and effectively reduces wastewater treatment costs.

After considering environmental and engineering aspects, finally, economical aspects need to be considered. In this combined process, the operating expenses include electricity, flocculant, and labor. The total installed capacity of the station is 123.85 kW, the operating power is 104.75 kW, and the power consumption is 1165.40 kW·h / d. The electricity cost is calculated at 0.80 yuan/[kW • h], then the electricity fee is 2.33 yuan / m^3. The reagents include H_2SO_4, NaOH, PAM, PAC, and H_2O_2. Under normal circumstances, the reagent fee is equivalent to 2.0 yuan/m^3. 3 wastewater stations need to be arranged. The staff adopts a three-shift system, and the salary is calculated at 2,000 yuan/[month person], equivalent to a labor cost of 0.5 yuan/m^3. The total operating cost is 4.96 RMB/m^3. In common, above mentioned combine process is 3E advisable which can satisfy environmental, engineering and economical aspects of the process.

Conclusion

Fine chemical manufacturing company, Ningxia producing fenfluramine, metalaxyl, difenoconazole, and other chemical synthetic pesticides and pesticide intermediates. In China, every enterprise need to meet the effluent water quality [GBT 31962-2015] Class B standard. In wastewater samples of Nangxia industry, high COD, high salt content, and high ammoniacal nitrogen were found. The biodegradability of wastewater is poor due to presence of higher concentration of pollutant. In this research, wastewater treatment process were combined by a mass pretreatment + biochemical treatment + advanced treatment process for further treatment. Biochemical treatment is based on hydrolysis and acidification, two-stage EGSB, and A/O process. The combined process has a good treatment effect, strong impact load resistance, and a high degree of automation as well. The final concentrations of effluent COD, NH_3-N and TDS were 376 mg / L, 34.74 mg / L, and 442 mg / L, and the removal rates were 99.9%, 81.3%, and 99.9%, respectively. Finally, the effluent water quality fluctuation range is small, and also it meets 3E analysis. This result meets the requirements of the enterprise to access the sewage pipe network of the park.

References

[1] Y. Liang, D. Wei, J. Hu, J. Zhang, Z. Liu, A. Li, and R. Li, Glyphosate and nutrients removal from simulated agricultural runoff in a pilot pyrrhotite constructed wetland, Water Res. 168 (2020) 115154. https://doi.org/10.1016/j.watres.2019.115154

[2] A. Serra-Clusellas, L. De Angelis, M. Beltramo, M. Bava, J. De Frankenberg, J. Vigliarolo, N. Di Giovanni, J.D. Stripeikis, J.A. Rengifo-Herrera, and M.M.F. de Cortalezzi, Glyphosate and AMPA removal from water by solar induced processes using low Fe[iii] or Fe[ii] concentrations, ENVIRONMENTAL SCIENCE-WATER RESEARCH & TECHNOLOGY 5 (2019) 1932-1942. https://doi.org/10.1039/C9EW00442D

[3] C. Penn, J. Gonzalez, M. Williams, D. Smith, and S. Livingston, The past, present, and future of blind inlets as a surface water best management practice, Crit. Rev. ENV. Sci. Tec. 50 (7) (2020) 743-768. https://doi.org/10.1080/10643389.2019.1642836

[4] A. Derbalah, R. Chidya, W. Jadoon, and H. Sakugawa, Temporal trends in organophosphorus pesticides use and concentrations in river water in Japan, and risk assessment, JOURNAL OF ENVIRONMENTAL SCIENCES-CHINA 79 (2019) 135-152. https://doi.org/10.1016/j.jes.2018.11.019

[5] K.S.G.C. Oliveira, R.M. Farinos, A.B. Veroli, and L.A.M. Ruotolo, Electrochemical incineration of glyphosate wastewater using three-dimensional

electrode, Environ. Technol. (2019) 1-12. https://doi.org/10.1080/09593330.2019.1625563

[6] B. Xing, H. Chen, and X. Zhang, Removal of organic phosphorus and formaldehyde in glyphosate wastewater by CWO and the lime-catalyzed formose reaction, Water Sci. Technol. 75 (2017) 1390-1398. https://doi.org/10.2166/wst.2017.006

[7] N. Tan, Z. Yang, X. Gong, Z. Wang, T. Fu, and Y. Liu, In situ generation of H2O2 using MWCNT-Al/O-2 system and possible application for glyphosate degradation, Sci. Total Environ. 650 (2019) 2567-2576. https://doi.org/10.1016/j.scitotenv.2018.09.353

[8] A.M.M. Cermak, I. Pavicic, and D. Zeljezic, Redox imbalance caused by pesticides: a review of OPENTOX-related research, Arh. Hig. Rada Toksiko. 69 (2018) 126-134. https://doi.org/10.2478/aiht-2018-69-3105

[9] T.S. Pinto Persch, R.N. Weimer, B.S. Freitas, and G.T. Oliveira, Metabolic parameters and oxidative balance in juvenile Rhamdia quelen exposed to rice paddy herbicides: Roundup [R], Primoleo [R], and Facet [R], Chemosphere 174 (2017) 98-109. https://doi.org/10.1016/j.chemosphere.2017.01.092

[10] R. Mesnage, G. Renney, G. Seralini, M. Ward, M.N. Antoniou, Multiomics reveal non-alcoholic fatty liver disease in rats following chronic exposure to an ultra-low dose of Roundup herbicide, Sci. Rep.-UK 7 (2017). https://doi.org/10.1038/srep39328

[11] T. Nam, P. Drogui, L.D. Tuan, S.L. Thanh, C.N. Hoai, Electrochemical degradation and mineralization of glyphosate herbicide, ENViron. Technol. 38 (2017) 2939-2948. https://doi.org/10.1080/09593330.2017.1284268

[12] Y. Li, H. Wang, L. Zhu, Z. He, R. Liu, Treatment of glyphosate wastewater and recovery of phosphorus by sodium hypochlorite oxidation-magnesium ammonium phosphate precipitation process, Environmental Protection of Chemical Industry 37 (2017) 627-631.

[13] T. Zheng, Y. Sun, Y. Lin, N. Wang, P. Wang, Study on preparation of microwave absorbing MnOx/Al_2O_3 adsorbent and degradation of adsorbed glyphosate in MW-UV system, Chem. Eng. J. 298 (2016) 68-74. https://doi.org/10.1016/j.cej.2016.03.143

[14] H. Lan, W. He, A. Wang, R. Liu, H. Liu, J. Qu, C.P. Huang, An activated carbon fiber cathode for the degradation of glyphosate in aqueous solutions by the Electro-Fenton mode: Optimal operational conditions and the deposition of iron on cathode on electrode reusability, Water Res. 105 (2016) 575-582. https://doi.org/10.1016/j.watres.2016.09.036

Chapter 7

Photocatalytic Degradation of Levofloxacin by Cu doped TiO_2 under Visible LED Light

K.S. Varma[1], V.G. Gandhi[1]*, R.J. Tayade[2], A.D. Shukla[1], B. Bharatiya[3], P.A. Joshi[1]

[1]Department of Chemical Engineering & Shah Schulman Center for Surface Science and Nanotechnology, Dharmsinh Desai University, College Road, Nadiad 387 001, Gujarat, India

[2]Discipline of Inorganic Materials and Catalysis, Central Salt and Marine Chemicals Research Institute, Council of Scientific and Industrial Research (CSIR), G. B. Marg, Bhavnagar-364002, Gujarat, India

[3]School of Chemistry, University of Bristol, England

*vggandhi.ch@ddu.ac.in

Abstract

Degradation performance of Cu-TiO_2 photocatalytic materials against a widely used antibiotic drug levofloxacin (LFX) was investigated under visible LED light source of 40 W. Cu-TiO_2 (0.25-1.0 wt%) nanomaterials are prepared through reverse micelle mediated modified sol-gel method. Characterization of synthesized Cu-TiO_2 samples are performed by XRD, UV-Vis, and DLS techniques. The doping of 0.5 wt% copper in TiO_2 shown lower crystallite size (5.79 nm) and visible light absorption characteristics with energy band gap of 2.84 eV. 0.5 wt% Cu-TiO_2 photocatalyst has shown significant LFX degradation of 75.5% with catalyst loading of 1 g/L and initial pollutant concentration of 50 mg/L.

Keywords

Cu-TiO_2, Doping, Photocatalyst, Degradation, Levofloxacin, Visible LED Light

Contents

1. Introduction

Water pollution through antibiotics is a huge challenge for industries, municipal bodies and environmental protection agencies. Based on chemical structure of antibiotics, it is categorized mainly into macrolides, tetracyclines, fluoroquinolones and sulphonamides [1]. Extensively used different fluoroquinolones (FQs) are detected in industrial and municipal wastewater as well as in surface and ground water resources in different countries of the world viz. Argentina, China, Egypt, India, Iran, Pakistan, and Spain [2-9]. Moreover, Levofloxacin (LFX) which is one of the widely used FQs class antibiotic for different bacterial infections is observed in wastewater due to its persistency to biodegradation [10-13]. Existence of LFX in water is toxic to cyanobacteria which is crucial member of nutrient cycle in aquatic ecosystem [14,15]. Conventional processes like flocculation and sedimentation are found ineffective towards LFX removal from drinking water treatment plant and removal efficiency of LFX by conventional activated sludge process is less than 50% in sewage wastewater treatment plant [16,17]. In vicinity of pharmaceutical industrial sector, significant concentration of LFX has been observed in water resources due to less upgradation of industrial wastewater treatment plant [6].
Hence, the role of advanced wastewater treatment technologies is becoming critically important to resolve the obstacle associated with LFX pollution in water resources. Adsorbents like carbon, Al_2O_3 and SiO_2 based nanomaterials have shown effective removal of levofloxacin [18-20]. However, the adsorption process only transferred the aqueous pollutants from one phase to another, which generated contaminants adsorbed solid residues [21]. In view of this, advanced oxidation processes (AOPs) are much explored due

Materials Research Forum LLC
https://doi.org/10.21741/9781644901397-7

to their effective degradation of various pollutants to addresses environmental issues related especially pharmaceutical and dye contaminants [22-26]. Amongst the various AOPs, nanomaterials assisted photocatalytic processes have shown promising results in degradation of LFX [27-30]. TiO_2 based nanomaterials are investigated for LFX and other FQs antibiotics degradation under UV, solar and visible light source due to their features like non-toxicity, chemical stability and high oxidizing capacity [31-39].

Selection of the light source is crucial in the photocatalytic process, which influences the degradation performance as well as energy consumption. Conventional UV light source like mercury lamps have shown limitations like high energy consumption, contains hazardous mercury and cooling water requirement for controlling the temperature inside the photocatalytic reactor [40]. Light emitting diodes (LEDs) as light source are shown to provide advantages such as longer life compared to conventional light source, less utilization of energy and also it is not harmful to the environment [41]. UV LED is employed by few researchers in photocatalytic reactors, but UV LED is considerably costlier than visible LED [42]. Moreover, Visible LED light sources are safer for operational personnel with respect to process safety, whereas UV light is harmful to human and caused skin diseases [43].

Wide band gap energy semiconductor material TiO_2 (anatase phase ~ 3.2 eV) has shown effective photo catalytic activity mainly under a UV light source. Also, TiO_2 material hardly absorb the solar light as the natural light source is comprised of less than 5% UV region [44]. For development of visible light active TiO_2 based nanaomaterials, doping of metal and non-metal, heterojunctions semiconductor and other strategies are utilized which altered the photocatalytic properties such as decreasing band gap energy and delaying the electron-hole pair recombination rate [45]. Metal doping is one of the most widely used strategies, which effectively shifting band gap edge towards visible light wavelength and also helps in charge carrier separation [46,47]. There are different metals such as Ag, Au, Fe, Ni, Zn and Cu which are reportedly utilized for development of visible light active metal doped TiO_2 nanomaterials [48-53]. Copper metal is cheaper and readily available as compared to costlier noble metals. It is observed that through electron trapping, copper dopant is reported to play a vital role in delaying rate of charge carrier recombination. Also, copper metal doping onto TiO_2 is shifting the band gap energy towards the visible light region [54-56].

Most commonly synthesize methods for metal doped TiO_2 nanomaterials are sol-gel and hydrothermal [57,58]. Synthesis of nanomaterials through reverse micelle mediated method provides better control over the morphology of nanomaterials. Synthesizing the nanomaterials with narrow particle size distribution (PSD) in the reverse micelle method

Materials Research Forum LLC
https://doi.org/10.21741/9781644901397-7

is more convenient compared to other methods as reverse micelle acts as nanodomains during the hydrolysis reaction [59,60].

Hence, in the present work, the reverse micelle mediated modified sol-gel method is used for synthesizing Cu-TiO_2 materials with different copper doping concentrations (0.25-1.0 wt%). Synthesized Cu-TiO_2 materials are utilized to degrade the environmentally harmful LFX antibiotic under visible light LED source in order to develop the sustainable photocatalyst based environmental remediation process.

2. Materials and methods

2.1 Materials and reagents

For synthesizing Cu doped TiO_2 materials, precursors titanium tetraisopropxide (TTIP) (> 98%) and copper (II) nitrate trihydrate extra pure (≥ 99 %) are purchased from Spectrochem Pvt. Ltd., Mumbai and Merck Specialities Pvt. Ltd., Mumbai respectively. Triton X-114 (laboratory grade) is purchased from Sigma-Aldrich Chemicals Pvt. Ltd., Bangalore. Hexane (fraction from petroleum) and Toluene (≥ 99.0 %) are procured from S D Fine-Chem Limited, Mumbai and Merck Specialities Pvt. Ltd., Mumbai separately. Pharmaceutical drug levofloxacin (LFX) is received from Rhombus pharma private limited, Gujarat.

2.2 Preparation of Cu-TiO_2 photocatalyst

Reverse micelle nanodomains mediated modified sol-gel method is utilized for synthesis of Cu doped TiO_2 materials. The molar ratio of solvent/surfactant/$CuNO_3$ aqueous solution/TTIP precursor is kept at 11/1/1/1. Solvent is consisted the hexane and toluene, which is taken in 7:3 volume ratio and triton x-114 is taken as surfactant. Reverse micelle is formed with intense mixing of surfactant and solvent mixture. For preparation of various (0.25-1.0 wt%) Cu doped TiO_2 materials, suitable concentration of $CuNO_3$ solution is added at very slow rate into the solvent-surfactant reverse micelle mixture. Swelling of reverse micelle is initiated due to addition of copper nitrate aqueous solution. In this hydrated reverse micelle solution, TTIP is added at slow rate, where hydrolysis is taken place inside the reverse micelle which act as a nanocarriers. At the end of the complete addition of TTIP, gel solution is reserved under stirring condition for around 15 H. To separate the gel from the solvent-surfactant mixture, it is centrifuged at 8000 rpm for 10 min. After gel is exposed to air for around six hours, drying is carried out at 130°C for 18 H in a hot air oven. Dried sample is calcined at 400°C for 4 H by using muffle furnace, where temperature increment rate is around 10°C/min. Samples 0.25 wt%, 0.5 wt%, 0.75

wt% and 1 wt% Cu-TiO_2 are designated as 0.25CuT, 0.5CuT, 0.75CuT and 1.0CuT respectively.

2.3 Characterization

For phase identification and measurement of crystallite size, XRD experiment is performed using Bruker AXS (D8 advance) X-ray powder diffractometer with Cu Kα radiation (λ = 1.5418 Å), where scanning of Cu-TiO_2 sample is in 2θ range of 10-85°. For determining light absorbance behavior of Cu-TiO_2 sample, Ultraviolet- Visible- Near IR (UV-Vis-NIR) spectroscopy is performed using Agilent Cary 5000i spectrophotometer. The average particle size and particle size distribution (PSD) of is analysed using dynamic light scattering (DLS) using Malvern particle size analyser NanoZS, where Cu-TiO_2 samples are dispersed in Millipore water and sonicated for around 30 min to isolate the agglomerated particles.

2.4 Photocatalytic degradation

For degradation of LFX (50 mg/L) under visible light irradiation, Philips 40 W LED is utilized as visible light source, where catalyst loading of Cu-TiO_2 samples are 1 g/L. Batch reactor volume is 350 ml and continuous aeration is provided through purging inside the photoreactor. Stirring condition of 300 rpm is kept for complete dispersion of photocatalyst in a levofloxacin aqueous solution. Sample is taken out at regular time interval and it is centrifuged at 9500 rpm for 10 min for separating Cu-TiO_2 nanomaterials. The absorption spectra of LFX is analysed in a UV-vis spectrophotometer. The rate of degradation is determined through change in λ_{max} of LFX (288 nm). Degradation is estimated using Eq. (1) as follow:

$$\text{Degradation (\%)} = [(C_0-C)/C_0] \times 100 \quad (1)$$

Where C_0 is initial concentration of LFX and C is concentration of LFX after visible light irradiation at any time t.

3. Results and discussion

3.1 Characterization of Cu-TiO_2 photocatalyst

X-ray diffraction profile for Cu doped TiO_2 samples calcined at 400°C are shown in Fig. 1. For TiO_2 anatase phase (denoted as A), diffraction peaks at 25.3°, 37.9°, 48.1°, 54.1°, 55.1° and 62.7° related to (101), (004), (200), (105), (211), and (201) orientations of TiO_2 anatase phase (ICDD No.01-089-4921) [61,62]. Fraction of anatase phase (X_A) in Cu-TiO_2

Materials Research Forum LLC
https://doi.org/10.21741/9781644901397-7

samples are calculated by $X_A = 1/(1 + 1.26\ I_R/I_A)$, peak intensities of anatase (101) and rutile (110) is corresponded to I_A and I_R respectively. Crystallite size is calculated by using Scherrer formula $D = 0.94\lambda/\beta\cos\theta$, D is corresponding to average crystallite size, λ is X-ray radiations' wavelength (1.5406Å), β is represented the full width at half maximum (FWHM) and θ is bragg angle [63]. Average crystallite size and anatase (%) are specified in Table 1, where 0.5CuT sample is exhibited lowest crystallite size. In higher Cu doped TiO_2 samples, crystallite size is increased, possibly due to enhancement of crystallization of TiO_2 [64]. Synthesized Cu doped TiO_2 samples are consisted nearly identical anatase phase in the range of 93-95%.

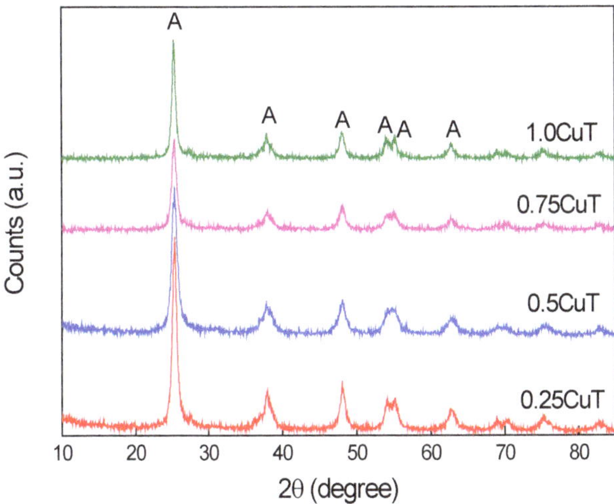

Figure 1. X-ray diffraction patterns of Cu-TiO_2 samples.

Absorbance profile of Cu-TiO_2 materials through UV-vis spectroscopy analysis are shown in Fig. 2a. It is observed that with increment in copper doping, absorbance edge is shifted towards higher wavelength. For determining band gap energy of the materials, tauc plot of light energy $(\alpha h\nu)^2$ vs. energy (hν) is plotted as shown on Fig. 2b, where it is found that band gap energy is decreased with increasing the Cu doping concentration. Values of band gap energy for synthesized samples are given in Table 1. Similar kind of observation is found for hydrothermally and sol-gel synthesized Cu doped TiO_2 nanomaterials, where declining trend is observed in band gap energy with increase in copper doping onto TiO_2 [65,66].

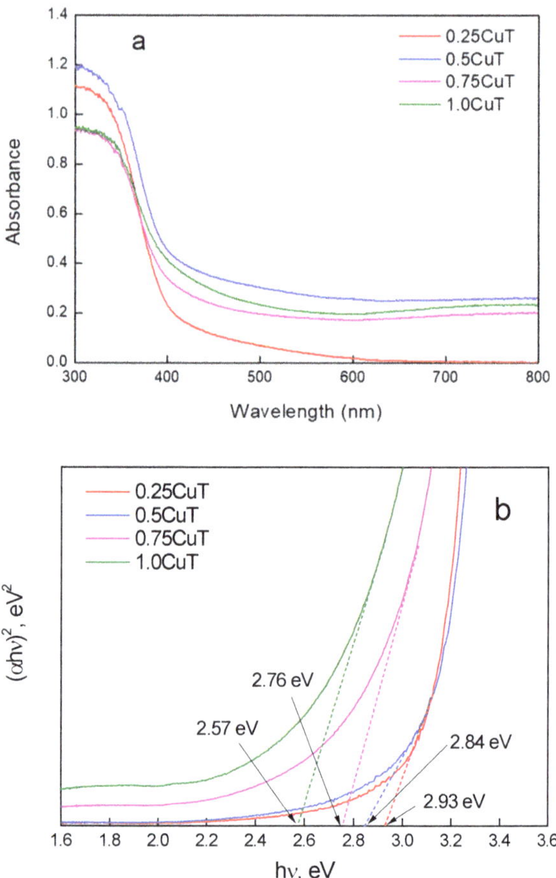

Figure 2. (a) Absorbance spectra profile of Cu-TiO_2 samples, (b) tauc plot.

Number average particle size of synthesized Cu-TiO_2 samples are analysed using DLS technique. It is found that particle size in the range of 50-70 nm for 0.25, 0.5 and 0.75 wt% Cu-TiO_2 samples as shown in Table 1. However, for 1.0CuT sample, particle size is relatively high, which may be due to high rate of agglomeration of nanomaterials in aqueous medium.

Materials Research Forum LLC
https://doi.org/10.21741/9781644901397-7

Table 1: Characteristics of Cu-TiO2 samples.

Nanomaterials	Average crystallite size (nm) [a]	Number average particle size (nm) [b]	Anatase (%)	Band gap energy (eV)
0.25CuT	8.10	55.0	93.99	2.93
0.5CuT	5.79	50.2	95.14	2.84
0.75CuT	8.22	70.8	94.11	2.76
1.0CuT	9.17	131.3	94.20	2.57

[a] using Scherrer formula; [b] using DLS technique

4. Degradation performance of Cu-TiO_2 towards LFX under visible LED light

To study the effect of Cu doping concentration on degradation performance under visible light LED, initial LFX concentration and photocatalyst concentration are kept constant as 50 mg/L and 1 g/L respectively. Also, to perform photolysis, effect of 40 W visible LED on degradation of LFX (50 mg/L) is checked, where only around 5% LFX is degraded under 6 H light irradiation. With doping of Cu onto TiO_2, band gap is tuned and it is reduced sufficiently to absorb the visible light wavelength. With using 0.5CuT material, more than 75% degradation of LFX is observed under visible light source in 6 H time span as shown in Fig. 3. Other Cu-TiO_2 samples are also shown good photocatalytical activity against LFX. This suggested that synthesized Cu doped TiO_2 nanomaterials are shown considerable degradation performance against antibiotic aqueous solution under visible light source. The photocatalytic degradation efficiency is observed in the order of 0.5CuT>0.25CuT>0.75CuT>1.0CuT.

Kinetics of degradation of LFX is simulated by Langmuir-Hinshelwood model as follow [67]:

$$-\ln (C/C_0) = kt \quad (2)$$

Where, C_0 is initial concentration of LFX, C is concentration of LFX after visible light photocatalytic degradation, t is irradiation time (min) and k is first order rate constant (min^{-1}).

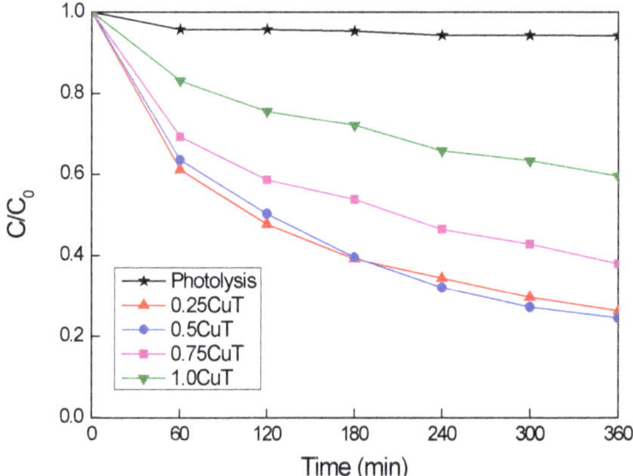

Figure 3. Degradation of LFX with and without presence of Cu-TiO_2 samples under 40W visible LED (initial LFX concentration = 50 mg/L, catalyst loading = 1 g/L).

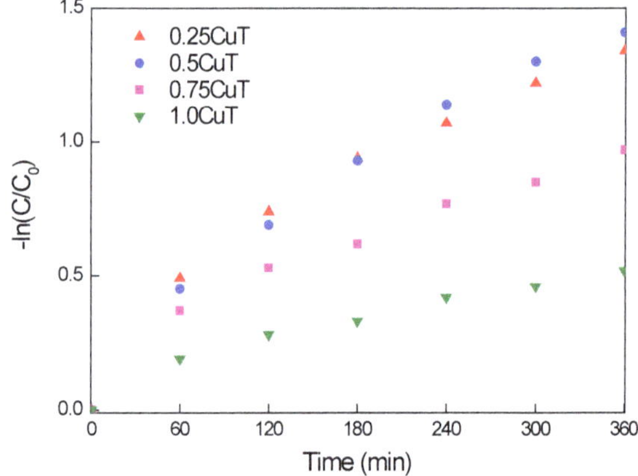

Figure 4. Kinetics of LFX degradation for Cu-TiO_2 samples (initial LFX concentration = 50 mg/L, catalyst loading = 1 g/L).

Materials Research Forum LLC
https://doi.org/10.21741/9781644901397-7

Table 2: First-order rate constant for LFX degradation with different Cu doped TiO_2 samples.

Nanomaterials	$k \times 10^3$ (min^{-1})	R^2
0.25CuT	4.25	0.97
0.5CuT	4.42	0.98
0.75CuT	3.01	0.97
1.0CuT	1.62	0.97

Kinetics for photocatalytic degradation of LFX using different Cu doped TiO_2 photocatalyst is shown in Fig. 4 and their rate constant values are given in Table 2. It is found that degradation performance is slightly enhanced, when copper doping concentration increases from 0.25 to 0.5 wt%. Moreover, it is found that 1.0CuT material has shown lowest photocatalytic activity, even though its band gap energy is low enough to absorb the significant photon energy from visible light source.

Conclusion

Cu-TiO_2 nanomaterials with different copper doping concentration is successfully synthesized through reverse micelle mediated modified sol-gel method. With increment in copper doping onto TiO_2, band gap wavelength is more shifted towards visible light region. Amongst the synthesized various Cu-TiO_2 materials, 0.5CuT is found to be having lower crystallite and particle size as well as the most efficient candidate for photocatalytic degradation of LFX under visible LED light, where more than 75% LFX degradation is taken place in 6 H time span.

Acknowledgments

Authors deeply acknowledge the Gujarat Pollution Control Board (GPCB), Gandhinagar, Gujarat (India) for providing Ph.D. fellowship (Grant ID: Ph.D./R&D/05/444437). The authors are grateful to Dharmsinh Desai University, Nadiad for providing essential research facilities and thankful to Dr. B. N. Suhagia, Dean, Faculty of Pharmacy, Dharmsinh Desai University for his valuable guidance in selection of pharmaceutical pollutants for photodegradation study.

References

[1] E. Etebu, I. Arikekpar, Antibiotics: Classification and mechanisms of action with emphasis on molecular perspectives, Int. J. Appl. Microbiol. Biotechnol. Res. 4 (2016), 90–101

[2] M. Boy-Roura, J. Mas-Pla, M. Petrovic, M. Gros, D. Soler, D. Brusi, A. Menció, Towards the understanding of antibiotic occurrence and transport in groundwater: Findings from the Baix Fluvià alluvial aquifer (NE Catalonia, Spain). Sci. Total Environ, 612 (2018) 1387–1406. https://doi.org/10.1016/j.scitotenv.2017.09.012

[3] M. Feng, X. Wang, J. Chen, R. Qu, Y. Sui, L. Cizmas, Z. Wang, V. K. Sharma, Degradation of fluoroquinolone antibiotics by ferrate(VI): Effects of water constituents and oxidized products. Water Res. 103 (2016) 48–57. https://doi.org/10.1016/j.watres.2016.07.014

[4] R. Mirzaei, M. Yunesian, S. Nasseri, M. Gholami, E. Jalilzadeh, S. Shoeibi, A. Mesdaghinia, Occurrence and fate of most prescribed antibiotics in different water environments of Tehran, Iran. Sci. Total Environ. 619–620 (2018) 446–459. https://doi.org/10.1016/j.scitotenv.2017.07.272

[5] P. K. Mutiyar, A. K. Mittal, Risk assessment of antibiotic residues in different water matrices in India: key issues and challenges. Environ. Sci. Pollut. Res. 21 (2014) 7723–7736. https://doi.org/10.1007/s11356-014-2702-5

[6] L. Riaz, T. Mahmood, A. Kamal, M. Shafqat, A. Rashid, Industrial release of fluoroquinolones (FQs) in the waste water bodies with their associated ecological risk in Pakistan. Environ. Toxicol. Pharmacol. 52 (2017) 14–20. https://doi.org/10.1016/j.etap.2017.03.002

[7] C. Teglia, F. Perez, N. Michlig, M. Repetti, H. Goicoechea, M. Culzoni, Occurrence, Distribution, and Ecological Risk of Fluoroquinolones in Rivers and Wastewaters. Environ. Toxicol. Chem. 38 (2019). https://doi.org/10.1002/etc.4532

[8] H. A. Younes, H. M. Mahmoud, M. M. Abdelrahman, H. F. Nassar, Seasonal occurrence, removal efficiency and associated ecological risk assessment of three antibiotics in a municipal wastewater treatment plant in Egypt. Environ. Nanotechnology, Monit. Manag. 12 (2019) 100239. https://doi.org/10.1016/j.enmm.2019.100239

[9] L. Zhang, L. Shen, S. Qin, J. Cui, Y. Liu, Quinolones antibiotics in the Baiyangdian Lake, China: Occurrence, distribution, predicted no-effect concentrations (PNECs) and ecological risks by three methods. Environ. Pollut. 256 (2020) 113458. https://doi.org/10.1016/j.envpol.2019.113458

[10] L. Birošová, T. Mackuľak, I. Bodík, J. Ryba, J. Škubák, R. Grabic, Pilot study of seasonal occurrence and distribution of antibiotics and drug resistant bacteria in wastewater treatment plants in Slovakia. Sci. Total Environ. 490 (2014) 440–444. https://doi.org/10.1016/j.scitotenv.2014.05.030

[11] G. Chen, X. Liu, D. Tartakevosky, M. Li, Risk assessment of three fluoroquinolone antibiotics in the groundwater recharge system. Ecotoxicol. Environ. Saf. 133 (2016) 18–24. https://doi.org/10.1016/j.ecoenv.2016.05.030

[12] A. Mahmood, H. Alhaideri, F. Hassan, Detection of Antibiotics in Drinking Water Treatment Plants in Baghdad City, Iraq. Adv. Public Heal. (2019) 1–10. https://doi.org/10.1155/2019/7851354

[13] A. Wang, H. Wang, H. Deng, S. Wang, W. Shi, Z. Yi, R. Qlu, K. Yan, Controllable synthesis of mesoporous manganese oxide microsphere efficient for photo-Fenton-like removal of fluoroquinolone antibiotics. Appl. Catal. B Environ. 248 (2019) 298–308. https://doi.org/10.1016/j.apcatb.2019.02.034

[14] M. Gonzalez-pleiter, S. Gonzalo, I. Rodea-Palomares, F. Leganes, R. Rosal, K. Boltes, E. Marco, F. Fernandez-Piñas, Toxicity of five antibiotics and their mixtures towards photosynthetic aquatic organisms: Implications for environmental risk assessment. Water Res. 47 (2013). https://doi.org/10.1016/j.watres.2013.01.020

[15] Z. Zhou, Z. Zhang, L. Feng, J. Zhang, Y. Li, T. Lu, H. Qian, Adverse effects of levofloxacin and oxytetracycline on aquatic microbial communities. Sci. Total Environ. 734 (2020) 139499. https://doi.org/10.1016/j.scitotenv.2020.13949

[16] L. P. Padhye, H. Yao, F. T. Kung'u, C. -H. Huang, Year-long evaluation on the occurrence and fate of pharmaceuticals, personal care products, and endocrine disrupting chemicals in an urban drinking water treatment plant. Water Res. 51 (2014) 266–276. https://doi.org/10.1016/j.watres.2013.10.070

[17] M. Yasojima, N. Nakada, K. Komori, Y. Suzuki, H. Tanaka, Occurrence of levofloxacin, clarithromycin and azithromycin in wastewater treatment plant in Japan. Water Sci. Technol. 53 (2006) 227–233. https://doi.org/10.2166/wst.2006.357

[18] M. H. Al-Jabari, S. Sulaiman, S. Ali, R. Barakat, A. Mubarak, S. A. Khan, Adsorption study of levofloxacin on reusable magnetic nanoparticles: Kinetics and antibacterial activity. J. Mol. Liq. 291 (2019) 111249. https://doi.org/10.1016/j.molliq.2019.111249

[19] S. S. Limbikai, N. A. Deshpande, R. M. Kulkarni, A. A. P. Khan, A. Khan, Kinetics and adsorption studies on the removal of levofloxacin using coconut coir charcoal impregnated with Al_2O_3 nanoparticles. Desalin. Water Treat. 57 (2016) 23918–23926. https://doi.org/10.1080/19443994.2016.1138330

[20] A. Ullah, M. Zahoor, S. Alam, R. Ullah, A. S. Alqahtani, H. M. Mahmood, Separation of levofloxacin from industry effluents using novel magnetic

nanocomposite and membranes hybrid processes. Biomed Res. Int. (2019) 5276841. https://doi.org/10.1155/2019/5276841

[21] V. Homem, L. Santos, Degradation and removal methods of antibiotics from aqueous matrices – A review. J. Environ. Manage. 92 (2011) 2304–2347. https://doi.org/10.1016/j.jenvman.2011.05.023

[22] B. Bethi, S. H. Sonawane, B. A. Bhanvase, S. P. Gumfekar, Nanomaterials-based advanced oxidation processes for wastewater treatment: A review. Chem. Eng. Process.- Process Intensif. 109 (2016) 178–189. https://doi.org/10.1016/j.cep.2016.08.016

[23] M. Klavarioti, D. Mantzavinos, D. Kassinos, Removal of residual pharmaceuticals from aqueous systems by advanced oxidation processes. Environ. Int. 35 (2009) 402–417. https://doi.org/10.1016/j.envint.2008.07.009

[24] R. Patel, T. Bhingradiya, A. Deshmukh, V. Gandhi, Response Surface Methodology for Optimization and Modeling of Photo-Degradation of Alizarin Cyanine Green and Acid Orange 7 Dyes Using UV/TiO_2 Process. Mater. Sci. Forum 855 (2016) 94–104. https://doi.org/10.4028/www.scientific.net/MSF.855.94

[25] J. Vyas, M. Mishra, V. Gandhi, Photocatalytic Degradation of Alizarin Cyanine Green G, Reactive Red 195 and Reactive Black 5 Using UV/TiO_2 Process. Mater. Sci. Forum 764 (2013) 284–292. https://doi.org/10.4028/www.scientific.net/MSF.764.284

[26] R. J. Tayade, W. K. Jo, Enhanced Photocatalytic Activity of TiO_2 Supported on Different Carbon Allotropes for Degradation of Pharmaceutical Organic Compounds, in: R. J. Tayade, V. Gandhi (Eds.), Photocatalytic nanomaterials for environmental applications, Materials Research Forum LLC, Millersville, PA 17551, USA, 2018, pp. 139-159.

[27] B. S. M. Al Balushi, F. Al Marzouqi, B. Al Wahaibi, A. T. Kuvarega, S. M. Z. Al Kindy, Y. Kim, R. Selvaraj, Hydrothermal synthesis of CdS sub-microspheres for photocatalytic degradation of pharmaceuticals. Appl. Surf. Sci. 457 (2018) 559–565. https://doi.org/10.1016/j.apsusc.2018.06.286

[28] Q. Chen, Y. Xin, X. Zhu, Au-Pd nanoparticles-decorated TiO_2 nanobelts for photocatalytic degradation of antibiotic levofloxacin in aqueous solution. Electrochim. Acta 186 (2015) 34–42. https://doi.org/10.1016/j.electacta.2015.10.095

[29] M. Kaur, A. Umar, S. K. Mehta, S. K. Kansal, Reduced graphene oxide-CdS heterostructure: An efficient fluorescent probe for the sensing of Ag(I) and sunset yellow and a visible-light responsive photocatalyst for the degradation of levofloxacin

drug in aqueous phase. Appl. Catal. B Environ. 245 (2019) 143–158. https://doi.org/10.1016/j.apcatb.2018.12.042

[30] H. Sun, P. Qin, Z. Wu, C. Liao, J. Guo, S. Luo, Y. Chai, Visible light-driven photocatalytic degradation of organic pollutants by a novel Ag_3VO_4/Ag_2CO_3 p–n heterojunction photocatalyst: Mechanistic insight and degradation pathways. J. Alloys Compd. 834 (2020) 155211. https://doi.org/10.1016/j.jallcom.2020.155211

[31] T. An, H. Yang, W. Song, G. Li, H. Luo, W. J. Cooper, Mechanistic Considerations for the Advanced Oxidation Treatment of Fluoroquinolone Pharmaceutical Compounds using TiO_2 Heterogeneous Catalysis. J. Phys. Chem. A 114 (2010) 2569–2575. https://doi.org/10.1021/jp911349y

[32] V. Bhatia, A. K. Ray, A. Dhir, Enhanced photocatalytic degradation of ofloxacin by co-doped titanium dioxide under solar irradiation. Sep. Purif. Technol. 161 (2016) 1–7. https://doi.org/10.1016/j.seppur.2016.01.028

[33] X. Van Doorslaer, P. M. Heynderickx, K. Demeestere, K. Debevere, H. Van Langenhove, J. Dewulf, TiO_2 mediated heterogeneous photocatalytic degradation of moxifloxacin: Operational variables and scavenger study. Appl. Catal. B Environ. 111–112 (2012) 150–156. https://doi.org/10.1016/j.apcatb.2011.09.029

[34] V. Gandhi, M. Mishra, P. A. Joshi, Titanium dioxide catalyzed photocatalytic degradation of carboxylic acids from waste water: A review. Mater. Sci. Forum 712 (2012) 175–189. https://doi.org/10.4028/www.scientific.net/MSF.712.175

[35] S. K. Kansal, P. Kundu, S. Sood, R. Lamba, A. Umar, S. K. Mehta, Photocatalytic degradation of the antibiotic levofloxacin using highly crystalline TiO_2 nanoparticles. New J. Chem. 38 (2014) 3220–3226. https://doi.org/10.1039/C3NJ01619F

[36] A. Kaur, D. B. Salunke, A. Umar, S. K. Mehta, A. S. K. Sinha, S. K. Kansal, Visible light driven photocatalytic degradation of fluoroquinolone levofloxacin drug using Ag_2O/TiO_2 quantum dots: a mechanistic study and degradation pathway. New J. Chem. 41 (2017) 12079–12090. https://doi.org/10.1039/C7NJ02053H

[37] X. Qu, J. Brame, Q. Li, P. J. J. Alvarez, Nanotechnology for a Safe and Sustainable Water Supply: Enabling Integrated Water Treatment and Reuse. Acc. Chem. Res. 46 (2013) 834–843. https://doi.org/10.1021/ar300029v

[38] M. A. Rauf, M. A. Meetani, S. Hisaindee, An overview on the photocatalytic degradation of azo dyes in the presence of TiO_2 doped with selective transition metals. Desalination 276 (2011) 13–27. https://doi.org/10.1016/j.desal.2011.03.071

[39] S. Sharma, A. Umar, S. Mehta, A. Ibhadon, S. Kansal, Solar light driven photocatalytic degradation of levofloxacin using TiO_2/Carbon-dots nanocomposite. New J. Chem. 42 (2018) https://doi.org/10.1039/C7NJ05118B

[40] K. Natarajan, P. Singh, H. C. Bajaj, R. J. Tayade, Facile synthesis of $TiO_2/ZnFe_2O_4$ nanocomposite by sol-gel auto combustion method for superior visible light photocatalytic efficiency. Korean J. Chem. Eng. 33 (2016) 1788–1798. https://doi.org/10.1007/s11814-016-0051-4

[41] W. -K. Jo, R. J. Tayade, New Generation Energy-Efficient Light Source for Photocatalysis: LEDs for Environmental Applications. Ind. Eng. Chem. Res. 53 (2014) 2073–2084. https://doi.org/10.1021/ie404176g

[42] K. Natarajan, H. Bajaj, R. Tayade, Synthesis Route Impact on $BiVO_4$ Nanoparticles and their Visible Light Photocatalytic Activity Under Green LED Irradiation. J. Nanosci. Nanotechnol. 19 (2019) 5100–5115. https://doi.org/10.1166/jnn.2019.16833

[43] R. Klein, R. Sayre, J. Dowdy, V. Werth, The risk of ultraviolet radiation exposure from indoor lamps in lupus erythematosus. Autoimmun. Rev. 8 (2009) 320–324. https://doi.org/10.1016/j.autrev.2008.10.003

[44] D. Chatterjee, S. Dasgupta, Visible light induced photocatalytic degradation of organic pollutants. J. Photochem. Photobiol. C Photochem. Rev. 6 (2005) 186–205. https://doi.org/10.1016/j.jphotochemrev.2005.09.001

[45] H. Dong, G. Zeng, L. Tang, C. Fan, C. Zhang, X. He, Y. He, An overview on limitations of TiO_2-based particles for photocatalytic degradation of organic pollutants and the corresponding countermeasures. Water Res. 79 (2015) 128–146. https://doi.org/10.1016/j.watres.2015.04.038

[46] T. Parangi, M. K. Mishra, Titania Nanoparticles as Modified Photocatalysts: A Review on Design and Development. Comments Inorg. Chem. 39 (2019) 90–126. https://doi.org/10.1080/02603594.2019.1592751

[47] K. S. Varma, R. J. Tayade, K. J. Shah, P. A. Joshi, , A. D. Shukla, , V. G. Gandhi, Photocatalytic degradation of pharmaceutical and pesticide compounds (PPCs) using doped TiO_2 nanomaterials: A review. Water-Energy Nexus 3 (2020) 46–61. https://doi.org/10.1016/j.wen.2020.03.008

[48] S. S. Boxi, S. Paria, Visible light induced enhanced photocatalytic degradation of organic pollutants in aqueous media using Ag doped hollow TiO_2 nanospheres. RSC Adv. 5 (2015) 37657–37668. https://doi.org/10.1039/C5RA03421C

[49] L. Devi, N. Kottam, B. Murthy, S. Kumar, Enhanced photocatalytic activity of transition metal ions Mn^{2+}, Ni^{2+} and Zn^{2+} doped polycrystalline titania for the degradation of Aniline Blue under UV/solar light. J. Mol. Catal. A Chem. 328 (2010) 44–52. https://doi.org/10.1016/j.molcata.2010.05.021

[50] V. Krishnakumar, S. Boobas, J. Jayaprakash, M. Rajaboopathi, B. Han, M. Louhi-Kultanen, Effect of Cu doping on TiO_2 nanoparticles and its photocatalytic activity under visible light. J. Mater. Sci. Mater. Electron. 27 (2016) 7438–7447. https://doi.org/10.1007/s10854-016-4720-1

[51] X. Lin, F. Rong, X. Ji, D. Fu, Visible light photocatalytic activity and Photoelectrochemical property of Fe-doped TiO_2 hollow spheres by sol–gel method. J. Sol-Gel Sci. Technol. 59 (2011) 283–289. https://doi.org/10.1007/s10971-011-2497-5

[52] M. Vega, M. Hinojosa Reyes, A. Hernandez-Ramírez, J. Guzmán Mar, V. Glez, L. Reyes, Visible light photocatalytic activity of sol–gel Ni-doped TiO_2 on p-arsanilic acid degradation. J. Sol-Gel Sci. Technol. 85 (2018) https://doi.org/10.1007/s10971-018-4579-0

[53] M. T. Yilleng, E. C. Gimba, G. I. Ndukwe, I. M. Bugaje, D. W. Rooney, H. G. Manyar, Batch to continuous photocatalytic degradation of phenol using TiO_2 and Au-Pd nanoparticles supported on TiO_2. J. Environ. Chem. Eng. 6 (2018) 6382–6389. https://doi.org/10.1016/j.jece.2018.09.048

[54] S. Mathew, P. Ganguly, S. Rhatigan, V. Kumaravel, C. Byrne, S. J. Hinder, J. Bartlett, M. Nolan, S. C. Pillai, Cu-Doped TiO_2: Visible Light Assisted Photocatalytic Antimicrobial Activity. Appl. Sci. 8 (2018) 2067. https://doi.org/10.3390/app8112067

[55] H. Nishikiori, T. Sato, S. Kubota, N. Tanaka, Y. Shimizu, T. Fujii, Preparation of Cu-doped TiO_2 via refluxing of alkoxide solution and its photocatalytic properties. Res. Chem. Intermed. 38 (2012) 595–613. https://doi.org/10.1007/s11164-011-0374-z

[56] R. S. K. Wong, J. Feng, X. Hu, P. L.Yue, Discoloration and Mineralization of Non-biodegradable Azo Dye Orange II by Copper-doped TiO_2 Nanocatalysts. J. Environ. Sci. Heal. Part A 39 (2004) 2583–2595. https://doi.org/10.1081/ESE-200027013

[57] M. Asilturk, F. Sayılkan, E. Arpaç, Effect of Fe^{3+} Ion Doping to TiO_2 on the Photocatalytic Degradation of Malachite Green Dye under UV and Vis-Irradiation. J. Photochem. Photobiol. A Chem. 203 (2009) 64–71. https://doi.org/10.1016/j.jphotochem.2008.12.021

[58] S. I. Mogal, V. G. Gandhi, M. Mishra, S. Tripathi, T. Shripathi, P. A. Joshi, , D. O. Shah, Single-Step Synthesis of Silver-Doped Titanium Dioxide: Influence of Silver on

Structural, Textural, and Photocatalytic Properties. Ind. Eng. Chem. Res. 53 (2014) 5749–5758. https://doi.org/10.1021/ie404230q

[59] M. Fernández-García, X. Wang, C. Belver, J. C. Hanson, J. A. Rodriguez, Anatase-TiO_2 Nanomaterials: Morphological/Size Dependence of the Crystallization and Phase Behavior Phenomena. J. Phys. Chem. C 111 (2007) 674–682. https://doi.org/10.1021/jp0656611i

[60] B. Richard, J. -L. Lemyre, A. M. Ritcey, Nanoparticle Size Control in Microemulsion Synthesis. Langmuir 33 (2017) 4748–4757. https://doi.org/10.1021/acs.langmuir.7b00773

[61] L. -F. Chiang, R. Doong, Cu–TiO_2 nanorods with enhanced ultraviolet- and visible-light photoactivity for bisphenol A degradation. J. Hazard. Mater. 277 (2014) 84–92. https://doi.org/10.1016/j.jhazmat.2014.01.047

[62] P. Singla, O. P. Pandey, K. Singh, Study of photocatalytic degradation of environmentally harmful phthalate esters using Ni-doped TiO_2 nanoparticles. Int. J. Environ. Sci. Technol. 13 (2016) 849–856. https://doi.org/10.1007/s13762-015-0909-8

[63] J. Huang, X. Guo, B. Wang, L. Li, M. Zhao, L. Dong, X. Liu, Y. Huang, Synthesis and Photocatalytic Activity of Mo-Doped TiO_2 Nanoparticles. J. Spectrosc. (2015) 681850. https://doi.org/10.1155/2015/681850

[64] C. -J. Lin, W. -T. Yang, Ordered mesostructured Cu-doped TiO_2 spheres as active visible-light-driven photocatalysts for degradation of paracetamol. Chem. Eng. J. 237 (2014)131–137. https://doi.org/10.1016/j.cej.2013.10.027

[65] R. Kamble, S. Mahajan, V. Puri, H. Shinde, P. K. M. Garadkar, Visible Light-Driven high Photocatalytic Activity of Cu-Doped TiO_2 Nanoparticles Synthesized by Hydrothermal Method. Mater. Sci. Res. India 15 (2018) 197–208. https://doi.org/10.13005/msri/150301

[66] N. Turkten, Z. Cinar, A. Tomruk, M. Bekbolet, Copper-doped TiO_2 photocatalysts: application to drinking water by humic matter degradation. Environ. Sci. Pollut. Res. 26 (2019) 36096–36106. https://doi.org/10.1007/s11356-019-04474-x

[67] M. Sarafraz, M. Sadeghi, A. Yazdanbakhsh, M. M. Amini, M. Sadani, A. Eslami, Enhanced photocatalytic degradation of ciprofloxacin by black Ti^{3+}/N-TiO_2 under visible LED light irradiation: Kinetic, energy consumption, degradation pathway, and toxicity assessment. Process Saf. Environ. Prot. 137 (2020) 261–272. https://doi.org/10.1016/j.psep.2020.02.030

Keyword Index

About the Editors

Dr. Kinjal J. Shah
Associate Professor,
College of Urban Construction,
Department of Municipal Engineering,
Nanjing Tech University (NTU),
Nanjing, China 218816.
M: +86 13072541186, Email: kjshah@njtech.edu.cn;
kinjalshah8@gmail.com

Dr. Kinjal J. Shah is an Associate Professor in the department of Municipal Engineering, Nanjing Tech University (NTU-China), he is also working as a visiting researcher at Carbon Cycle Research Center (CCRC), National Taiwan University (NTU-Taiwan). He completed his Ph.D. in the field of applied science from Graduate Institute of Applied Science, National Taiwan University of Science and Technology (NTUST), Taiwan in 2015. He has started his research carrier in 2009 from Shah Schulman Center for Surface Science and Nanotechnology (SSCSSNT), DDU, India. During his last 10 years of research carrier, he has published more than 30 SCI papers in international repute journals, edited book on "Advances in Wastewater Treatment I" and 3 book chapters to his credit. He has received leading Young Scientist award from Society of Polymer Science Japan in 2019, Japan. One of his research was nominated by ENI award, 2019, in category of "Advance Environmental Solutions". In addition, he has received many domestic and international awards from different societies and institutional bodies. He has been invited as lead speaker to many international accredited conferences.

His current interests are in the field of green chemistry and nano technology for sustainable environment. At present, his lab is developing technologies for advanced gas adsorption, and water purification technologies. He is serving as associate editorial board member in "Current Analytical Chemistry" journal, Bentham Science.

Dr. Vimal G. Gandhi
Associate Professor,
Department of Chemical Engineering,
Dharmsinh Desai University (DDU),
Nadiad-387 001 (Gujarat) India.
Email: vggandhi.ch@ddu.ac.in

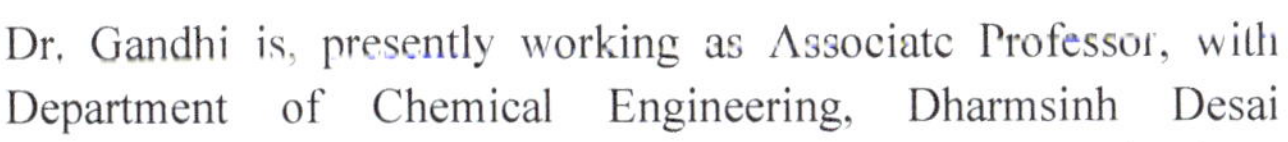

Dr. Gandhi is, presently working as Associate Professor, with Department of Chemical Engineering, Dharmsinh Desai University since last 20 years. He received his Ph.D. in the area of Application of nanotechnology in Environmental Engineering from Dharmsinh Desai University in 2011. He is serving as an Independent Director of BEIL Infrastructure Limited and Enviro Technology Limited (ETL) in Ankleshwar, Gujarat, India.

In the field of research and consultancy, He has guided several graduates and undergraduate students for their project work. He has more than 20 publications/presentation in International/National reputed journals and conferences to his credit. He edited books on -"Photocatalytic nanomaterials for Environmental Applications" and "Advances in Wastewater Treatment I" published by Material Research Forum, LLC- USA. His current interests are in the field of environmental engineering and synthesis of nanomaterials. He also organized various training programme for chemical industries in Gujarat including GNFC, PI Industries, Huntsman, Transpek Silox etc.

www.ingramcontent.com/pod-product-compliance
Ingram Content Group UK Ltd.
Pitfield, Milton Keynes, MK11 3LW, UK
UKHW021830270726
14058UKWH00001B/80